T0213133

Lecture Notes in Computer Science 1116

Edited by G. Goos, J. Hartmanis and J. van Leeuwen

Advisory Board: W. Brauer D. Gries J. Stoer

Springer
Berlin
Heidelberg
New York
Barcelona
Budapest
Hong Kong
London
Milan
Paris
Santa Clara
Singapore
Tokyo

Jane Hall   (Ed.)

Management of Telecommunication Systems and Services

Modelling and Implementing TMN-Based Multi-Domain Management

Springer

Series Editors

Gerhard Goos, Karlsruhe University, Germany

Juris Hartmanis, Cornell University, NY, USA

Jan van Leeuwen, Utrecht University, The Netherlands

Volume Editor

Jane Hall
GMD-Fokus
Hardenbergplatz 2, D-10623 Berlin, Germany
hall@fokus.gmd.de

Cataloging-in-Publication data applied for

Die Deutsche Bibliothek - CIP-Einheitsaufnahme

Management of telecommunication systems and services :
modelling and implementing TMN-based multi-domain
management / Jane Hall (ed.). - Berlin ; Heidelberg ; New
York ; Barcelona ; Budapest ; Hong Kong ; London ; Milan ;
Paris ; Santa Clara ; Singapore ; Tokyo : Springer, 1996
  (Lecture notes in computer science ; Vol. 1116)
  ISBN 3-540-61578-4
NE: Hall, Jane [Hrsg.]; GT

CR Subject Classification (1991): C.2, D.2, H.5, J.1, K.6, K.4.1, K.5.2

ISSN 0302-9743
ISBN 3-540-61578-4 Springer-Verlag Berlin Heidelberg New York

This work is subject to copyright. All rights are reserved, whether the whole or part of the material is
concerned, specifically the rights of translation, reprinting, re-use of illustrations, recitation, broadcasting,
reproduction on microfilms or in any other way, and storage in data banks. Duplication of this publication
or parts thereof is permitted only under the provisions of the German Copyright Law of September 9, 1965,
in its current version, and permission for use must always be obtained from Springer -Verlag. Violations are
liable for prosecution under the German Copyright Law.

© Springer-Verlag Berlin Heidelberg 1996
Printed in Germany

Typesetting: Camera-ready by author
SPIN 10513403   06/3142 - 5 4 3 2 1 0   Printed on acid-free paper

Foreword

With the advent of broadband communications, new generations of communications systems and services are being conceived, responding to the ever growing demands of the market place. Additionally, under the new regulatory framework being established in Europe, new transnational service providers and network operators are emerging.

In parallel, effective global communications are becoming an essential enabler for successful business enterprises. With such importance being placed on communications, the management of these communication assets (the systems which provide the networks and the services) becomes a key issue both for the providers and the user.

Recognising this importance, the European research programme RACE has supported a number of projects in the area of communications management.

The challenge facing communications management lies in the complexity of the communications environment, comprising as it does a multiplicity of network technologies and competing and cooperating providers. The work done within the framework of TMN has laid the foundation for handling this complexity by recognising different domains of management, where a domain represents either an ownership domain or a particular network technology domain.

One of the RACE projects, PREPARE, has successfully investigated, developed, and demonstrated solutions for end-to-end management of communication systems and services, concentrating particularly on interworking between different domains of management. The project has adopted a sound approach to integrate an extensive set of hardware and software elements in order to provide a solution for inter-domain management, based on a complex scenario of heterogeneous private and public networks. The excellent results obtained by the project have been acknowledged by a diverse set of actors, ranging from standards bodies, manufacturers, operators, service providers, to academia.

To provide a wider dissemination of the topic, the PREPARE project team has written this book, with the goal of presenting in a coherent, integrated, and readable form the issues addressed by the project, the motivation for the work carried out, and the key results obtained. The results described in this book constitute a major contribution to the field of communications management which will be reflected in a better quality of service for the customer and a lower cost of ownership for the providers. As the name of the project suggests, the results of the PREPARE project enable European actors to become better *prepared* to address the challenges of an open market for the provision of communication services.

November 1995

Mario Campolargo

Project Line Coordinator

European Commission - Programme RACE

Preface

The telecommunications industry is changing rapidly. With the advent of broadband networks the worldwide interconnection of distributed system components becomes feasible. The challenges arising for service management from this environment, the issues involved in providing cooperative management between different organisations and their management systems in order to manage advanced teleservices on an end-to-end basis, and the approach adopted by the European RACE II research project PREPARE project in designing and implementing prototype management systems over real testbeds are the subject of this book.

In the future, cooperative service management needs to be given particular attention if the capabilities of broadband interconnection networks are to be utilised to their full extent for the realisation of the global information society. There is at present little information available on the relevant issues and even less on implementation experience. Prospective readers of this book include those concerned with the design, development, delivery, and maintenance of telecommunications services in the global service market as well as anyone with a general interest in multi-domain management issues.

The PREPARE project (1992-95) investigated the issues likely to arise in a future liberalised telecommunications market when managing services across networks belonging to different organisations. A specific objective was to verify this work by deploying and demonstrating systems managing telecommunications services and the broadband network testbeds over which they operate in a realistic multi-vendor multi-network environment. PREPARE used existing standards as far as possible and enhanced them where needed. It adopted the Telecommunications Management Network (TMN) framework [M.3010] for designing and implementing its management systems due to TMN's widespread recognition and acceptance by public network operators and transport network service providers. Standards are not always easy to interpret and do not necessarily cover every aspect required and so the work in PREPARE could, by testing many of the ideas in real demonstrators, show how the relevant management architecture and computational and information models should be designed to support the management of teleservices on an end-to-end basis in an open service market. PREPARE has not only implemented management systems based on the standards but as a result of its experiences in this area has contributed extensions to them and indicated to standards bodies and other practitioners the real problems involved and how to tackle them.

Structure of the Book

This book has been structured so that the reader obtains an understanding of the problem area and general information about the PREPARE project first. This is followed by more specific details about how the project proceeded in its work and the results that it attained in implementing management systems for managing the networks and services of its testbeds. A brief overview of the structure is given below.

Chapter 1 introduces the developments leading to the global information society and the characteristics of this environment that are creating new challenges for telecommunications management. The main problems that have to be resolved in order to achieve multi-domain service management are discussed in order to highlight the issues in the PREPARE project's area of investigation.

In chapter 2 the background to the PREPARE project and the initial context in which the project was carried through is discussed. This relates to the RACE framework and to service management work in preparation for integrated broadband communications (IBC) in Europe. The aims of the project, its approach, its testbed configurations, and the management architectures it developed are described so that the reader obtains a good basis for understanding the work presented in the following chapters.

Chapter 3 presents the PREPARE development methodology adopted in the design phase in order to proceed from the conceptual work to a functioning system. The implementation of a multi-domain management system is not a trivial task and the various stages in the process that are discussed in this chapter show how the modelling work can be realised in an actual implementation.

Chapter 4 introduces multi-domain service management using examples from the PREPARE implementation work. This chapter provides an overview of what was actually managed in PREPARE with the examples showing in some detail how specific services were managed on an end-to-end basis. The services selected comprise an ATM Virtual Path service, a Virtual Private Network service, a Multimedia Mail service, and a Multimedia Conferencing service. Each service is discussed using a common approach. First a general overview of the service is given. Then the stakeholders for the service are introduced and the resources that are relevant to the service and its management are discussed. The management functionality is presented via the roles involved in the service and their responsibilities and obligations that require certain management functions to be carried out. As a result of the management function requirements, the information model, computational model, and information flows for each service are described in order to show how the management functionality was designed for each service. The interdependencies and interactions between the services are highlighted in order to demonstrate the complexities of managing multiple services in a multi-vendor multi-network environment.

During the period of the PREPARE project's work, promising developments were taking place in other areas of distributed computing that could usefully be investigated by those about to embark on developing management systems for the open service market. Chapter 5 evaluates these developments and their significance for future work on inter-domain service management. The particular relevance of each development to inter-domain management is emphasised in order to show what benefits each specific development can contribute to the area.

Chapter 6 provides a summary of the work undertaken by PREPARE and its achievements as well as a discussion of issues that are still open and requiring further investigation. This includes an evaluation of the TMN principles as experienced by PREPARE in its work, a view on how management in this area could evolve, and an overview of further work being carried out by the PREPARE partners.

The appendices provide additional information on certain aspects of the work referred to in the book. They are intended for readers who would like to know more about these aspects and aim to give a short introduction to the relevant concepts. Appendix A surveys the main principles behind the TMN recommendations. Readers less familiar with the TMN concepts are strongly encouraged to read this appendix as the work discussed in this book is based upon the TMN principles and a basic understanding of them is assumed. Appendix B introduces the main ideas behind OSI Systems Management, focusing on the concepts relevant to the TMN principles. Appendix C provides details of the Inter-Domain Management Information Service, which was developed in the PREPARE project in order to provide a global repository for management information required in a multi-domain environment. Appendix D presents the work that PREPARE undertook in investigating how the IN standards work needs to develop to prepare for the challenges of inter-domain service management in the global service market. Having gained so much experience from its implementation work, PREPARE developed architectural solutions for managing IN networks and services which are described in this appendix. Appendix E provides examples of the information models developed in PREPARE for managing the public network ATM service and the PREPARE VPN service.

This book is based on the work of the PREPARE project and so published papers are not individually referenced in the text. Papers published about the work undertaken in PREPARE on inter-domain management are listed separately as a guide to further information about the project's work. The glossary contains a useful list of the most important terms and concepts together with definitions of how they are used in the book. It explains how a particular term is used in PREPARE and how it is to be understood in the context of the PREPARE work. This is important for readers who may have experience of some of the terms in the book from other contexts.

The PREPARE Consortium

The PREPARE project was part of the European RACE II research programme. The work of the project was carried out between 1992 and 1995. The following organisations were members of the PREPARE consortium throughout this period:

- Broadcom Eireann Research Ltd, Dublin, Ireland.
- DSC Communications A/S, Copenhagen, Denmark (formerly NKT Elektronik A/S).
- L.M. Ericsson A/S, Copenhagen, Denmark.
- GMD-Fokus, Berlin, Germany.
- Compagnie IBM France S.A., La Gaude, France; IBM European Networking Center, Heidelberg, Germany.
- MARBEN, S.A. (SLIGOS Group), Agence Sud-Est, France.
- Tele Danmark A/S, Copenhagen, Denmark.
- University College London, United Kingdom.

Acknowledgements

PREPARE was a four-year research project involving many individuals who contributed in a variety of ways: motivating, reminding, organising, discussing, writing, criticising, designing, developing, implementing, testing, debugging, and publicising, to name just a few. The project would not have achieved the results it did without these enthusiastic contributions from every individual listed below. In addition to the considerable work required to implement PREPARE's testbeds, it was decided to produce a book to report on the work which, in turn, caused us to think critically about what we had done so far and to develop our ideas further. This book would never have come to fruition without the dedication of a few individuals who were concerned to see the results of PREPARE adequately recorded. They organised others to write for the book and contributed much input themselves. So although this book reflects the work of every member of the project, it could not have been produced without the particular efforts of Lennart H. Bjerring (L.M. Ericsson), David Lewis (University College London), Michel Louis (Marben), and Jürgen M. Schneider (IBM). To them, of course, the biggest thanks of all for such unwavering support and sheer hard work.

We should like to thank the reviewers for producing such perceptive and constructive comments and suggestions. We are grateful to GMD-Fokus for allowing the time to work on producing this book although it clashed with many other priorities, and specifically to Michael Tschichholz from GMD-Fokus who constantly supported and motivated the book project. Finally, as editor I wish to express my gratitude to David Crawford from GMD-Fokus who provided so much of the editing support essential in producing such a book.

Contributors

This book is the result of the work of many people. The following members of the above-mentioned organisations contributed to this book via their work in the PREPARE project.

Martin W. Andersen	Peter Gjerløv	Michael Klotz
Per K. Andersen	Kim Gormsen	Sven Krause
Søren Andersen	Lars Gredal	Bo Larsen
Philippe Bailet	Carsten Gyrn	Peter Levinsky
Alan Bartroff	Jane Hall	David Lewis
Michael van Bekkum	Per F. Hansen	Birgitte Lønvig
Jesper Birch	Uffe Harksen	Claus Lorenzen
Lennart H. Bjerring	David Hayes	Michel Louis
Ralf Bracht	Annemette Høgsbro	Carsten Lübbecke
Mark W. Burke	Jesper Holst	Rene S. Lund
Ingo Busse	Eamonn Howe	John F. MacDonald
Lars G. Christensen	Shane Hurley	Andreas Magnussen
James Cowan	Anne Hutton	Detlev Matthes
Jon Crowcroft	Holger Janssen	Flávio Morais de
Alina DaCruz	Mogens N. Jensen	Assis Silva
Andreas Dittrich	Xiaoping Johannsen	Jens D. Mouritzsen
Jean-Marc Djian	Henrik N. Jørgensen	Al Mullery
William Donnelly	Christian Kaas-	Bente H. Nielsen
Kim C. Dyresberg	Petersen	Carsten M. Nielsen
Claus P. Ek	Gautam Kar	Catherine G. Nielsen
Ann Fitzgerald	Peter Kirstein	Jacob V. Nielsen
Gabriele Gahse	Wolfram Kisker	Niels Q. Nielsen

Ove Nielsen
Per M. Nielsen
Peter S. Nielsen
Hans E. Nicolaysen
Henrik Pallisgørd
Stephan Paschke
Georg L. Pedersen
Svend M. Pedersen
Martin Pfeiler
Sonny Rasmussen
Alexander Richter
Knulp Rittmeyer
Pierre Rodier
Mehran Roshandel

Christian Rothe
Rong Shi
Oliver Schittko
Johanne Schmidt
Jürgen M. Schneider
Michael Sidenius
Morten Skov
Michael Slevin
Jens H. Sørensen
Lars B. Sørensen
Søren Sørensen
Jörg Strelow
Lars B. Svenningsen
Per H.Thomsen

Ingimundur H. Thorarensen
Thanassis Tiropanis
David Tracy
Ib Troldborg
Michael Tschichholz
Peter Viereck
Per Vorm
Susanne Waßerroth
Marcus Wittig
Olaf Zimmerman
Georg Zörntlein
Peter Zøylner

List of Tables

Table of Contents

List of Figures

1 Introduction

As the structure of the telecommunications industry moves towards that of a global service-driven market the role of telecommunications service developers will undergo a marked change. They will no longer be building services and management services for a single network but must utilise open interfaces to develop services over heterogeneous networks and platforms. Significantly, components that deliver and manage these services will not just be under the administrative control of the service providers, but of the customer and the providers of other constituent services also. The fact that resources have different owners must be taken into account when designing services and their management for this market. Multi-domain issues will therefore be playing a vital role in this environment.

In this book a *domain* is associated with an organisation. The set of resources, both virtual and real, belonging to an organisation comprises a management administrative domain, which is defined as a "management domain where the managed objects in the domain are all under the responsibility of one and only one administrative authority" [X.701]. As the majority of participants concerned with managing the global service market represent organisations, administrative domains provide a suitable basis for structuring management systems in this environment. Such a structure enables the autonomy of the participants to be preserved when managing end-to-end services via interactions between the individual domains. Other work on management domains has used the concept of domain to provide the framework for partitioning management responsibility and determining management policies for groups of objects in a large distributed system, for example in such projects as SysMan and IDSM [Sloman], Domino [SloTwi], and DOMAINS [Alpers]. The use of organisation to define a domain fits the requirements of the telecommunications environment and does not exclude the use of domains to further classify management functionality and define access to objects within a particular policy domain. We have not, however, used policy domains in this book.

We distinguish in this book between intra-domain, inter-domain, and multi-domaint of resources locally within a domain are termed *intra-domain* management. A domain has its own management system which is responsible for managing such resources locally, and how it manages these resources is not visible from outside the domain. *Inter-domain* management refers to the management activities across a management interface between any two separate domains. The management activities at this interface include the exchange of management information and cooperation to support the management of a service across the interface. These activities must be understood by each of the domains' management systems and adhere to a set of agreed constraints. The management information required to support inter-domain management must be made visible at each domain's interface and be accessible externally. This is a key interface and is the subject of much of the management work described in this book. *Multi-domain* management refers to the management activities required for the end-to-end management of services spanning several domains and comprises the management activities at all the inter-domain management interfaces involved in the management activities for any one service.

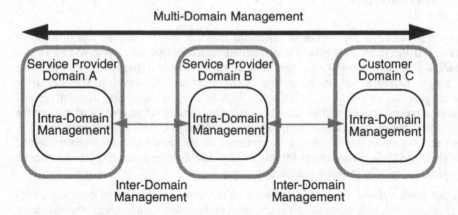

Figure 1.1: Intra-domain, inter-domain, and multi-domain management

This book is concerned with multi-domain management, particularly service management, within the Telecommunications Network Management (TMN) framework and with the work undertaken by the European research project PREPARE in this area. In a global advanced service environment the demands for flexible and transparent end-to-end services over large geographical distances, spanning several heterogeneous private and public networks, will affect many different administrative domains and require efficient cooperation of distributed management components across inter-domain boundaries. Service management on an end-to-end basis requires a level of integration that makes high demands on all aspects of management and so multi-domain management crystallises many of the problems confronting management of the global service market. Traditionally, both the telecommunications and computing worlds were based on the assumption of a single domain and this assumption determined the management for such a domain. However, the challenges of the global service environment extend beyond the capabilities of the traditional management infrastructure and management can no longer be tailored to suit only one domain's requirements.

Multi-domain management is mainly concerned with the question of how several autonomous domains can cooperate to provide end-to-end management so that service providers can deliver their services to customers and users efficiently and at the required quality level. It also requires an understanding of intra-domain capabilities in order to integrate them with the inter-domain capabilities at the domain interface. The key issue is the interoperability between the management systems and how to coordinate the management functionality at the border between the systems. Support for designing and implementing cooperative management systems providing an integrated approach to end-to-end management that meets the challenges created by technological and liberalisation developments is required, and it is exactly this that has been lacking both in the approaches to management and in existing management standards. The PREPARE project has gained valuable experience in designing, developing and demonstrating management systems for managing advanced telecommunications services crossing several domains, which is expected to provide useful knowledge about developing multi-domain management services in an advanced service environment.

The rest of this chapter discusses first the characteristics of the global service market which render multi-domain management necessary, and then the main issues that have to be considered when developing and implementing management systems for such an environment.

1.1   Towards the Twenty-First Century

> "Today we can sit anywhere at all on this planet and access the whole of human experience on a computer screen." [Gaarder]

> "The future is not a single national voice network with limited connection to the outside world. It will be a mass of interconnecting networks, under many different ownerships, of different geographical spreads, offering voice, image, text and data services from which the customer can choose quickly and easily." Tom McKinlay, DG XIII, European Commission [DeBony]

Current developments in networking and service engineering are providing the direction for the information-oriented, communications-based society of the twenty-first century in which users will have at their disposal an immense choice of services enabling them to create, process, store, retrieve, and transmit any kind of information in any medium anywhere. The merging of the telecommunications, computing, information, and entertainment industries that we are now witnessing is laying the foundation stone for this service-driven global information society where multimedia services over high speed gigabit (even terabit) networks will be available to both computer-literate and non-technical users not only from powerful desktop workstations with video interactive capabilities but also via a variety of other equipment, including set-top boxes and mobile equipment [Bangemann] [GII].

Worldwide interoperability of high speed networks with a wide range of services, including interactive and multimedia services both for residential and business purposes, will provide universal access to users who will use the services over these networks in the same way as applications on a single computer or local area network (LAN) are used today. Users will not know about interworking, from their perspective there is only *one* network, because for them it is the service and its use that is of interest, and not the underlying infrastructure. Where the users are located when using these services will be irrelevant as it will be possible to access the services throughout the world from both static and mobile equipment. The distinction between the use of services for business and personal use will also become less important as users at home are able to set up their own services to retrieve information easily. Users will be very different from those of today, they will have grown up in an environment where computers, communications, and multimedia technology are an integral part of their lifestyle. These users will be familiar with information technology, they will use it with confidence and will have high expectations about future developments [Cooper].

Currently, individual initiatives and pilot projects have created broadband islands for demonstrating the feasibility of the infrastructure for advanced service usage. These islands are being connected so that they will merge into one landscape where integrated networks support customised innovative and flexible multimedia services in a competitive arena. The major developments that are contributing to the global information society of the twenty-first century and the challenge of such developments to current management approaches are outlined below.

1.1.1   Technological Trends

A major factor in the development of the global information society is the widespread adoption of ATM (Asynchronous Transfer Mode) technology as the universal transmission technology [Vetter]. With transmission rates in a wide range from 16 Kbps up to 150 Mbps and improved price/performance ratios, integrated broadband communications over ATM can support the transmission of data streams carrying audio, video, and normal data over the same digital infrastructure. The advantages which an ATM-based broadband infrastructure offers with regard to cost-efficient and reliable global communications are the main reasons for the fast adoption of this technology throughout the communications industry. In both public and private network environments, non-integrated networks and services are being successively replaced by ATM-based broadband solutions [Snelling]. In addition, developments in technologies supporting mobile communications, such as satellite and cellular radio, are enabling the provision of reliable communications globally. A huge increase in network capacity is thus taking place, making available vast amounts of transmission, processing, and storage capabilities using bandwidth from a variety of providers. These capabilities will be introduced progressively and will favour the development of new distributed services.

At the same time personal computers, workstations, television-like devices, mobile equipment, powerful storage capabilities, and improved compression techniques to support highly demanding services are being developed to make effective use of the available bandwidth. Multimedia workstations able to generate, compress, and display streams of digital video and audio will lead to increased use of advanced services and in turn create demand for higher bandwidth. The high speed broadband networking will support this development so that video, audio, image, and data can be integrated and treated by users as available equally for any of the services they choose to use. At this point almost all imaginable communications and teleservices including multi-channel high-quality video will be available through a common broadband infrastructure, making the metaphor of the universal communications wall socket become true.

Sophisticated advanced services that are able to take advantage of these infrastructure developments are beginning to penetrate all areas of life, including publishing, education and training, medicine, and entertainment. There will not only be more services, but also a greater variety, both low-cost, high-volume services as well as high-quality customised services, including desktop video conferencing, the distributed office and teleworking, video-on-demand, services for those with special needs, and services for the personal home market, including teleshopping, telebanking, and home entertainment. Services will support the use of electronic transactions for a wide variety of applications, especially in societies which are culturally more open to such technological developments. The use of such services will be the driving force behind constant change, leading to new services not yet anticipated or envisaged which, in turn, will need to be rapidly created, deployed, and customised.

1.1.2   Regulatory and Business Trends

The technological changes affecting networks and services are being accompanied by an opening up of telecommunications markets to greater competition. The Open Network Provision (ONP) initiative in Europe [90/387/EEC] and the Open Network Architecture (ONA) initiative in the USA [Basu] are breaking up the traditional telecommunications monopolies by promoting a liberalised telecommunications environment in order to create competition and growth in the use of telecommunications services. Service providers must be given fair access to the underlying communications infrastructure, so enabling customers to purchase services from a variety of service provider organisations rather than from the restricted range of monopoly providers as in the past. This is leading to a competitive, aggressive environment with new providers without networks entering the market to offer services which before only monopoly operators could provide and forcing all service providers to look to the rapidly expanding global market in order to survive. New business niches and specialisations will be established and the number and range of services on offer will increase considerably, with traditional network carriers also offering value-added services in competition with other value-added service providers. This results in a greater choice and increased influence for customers as they can choose the services that best meet their requirements. Even more important in this context is the fact that the equal usage and supply provisions of ONP and ONA make the development and provision of effective multi-domain management essential as competing service providers can only operate under the same conditions when they have not only access to the underlying networking infrastructure but also to equal management capabilities over it.

As well as the technological and regulatory changes, general business trends are also having an effect on the telecommunications service market. Business has become more international and increasingly companies are having to operate on a global scale, either through mergers or joint agreements, which leads to greater demands for globally provided services regardless of the number of network operators involved in provisioning the infrastructure and bearer services. Global markets are reducing the significance of national boundaries as services must be available everywhere in the world on a 24 hour per day basis. Greater competition and the globalisation of business is leading to the development of new business practices requiring tight interorganisation relationships such as just-in-time production based on crucial customer-provider communication (for example, supported by electronic data interchange) or cooperative research and development demanding efficient dialogue between the partners, such as that supported by multimedia groupware tools.

At the same time, organisations are becoming more dependent on information and communications technologies to run their core business. The use of networks and distributed systems has increased dramatically in the last decade. Firms have developed their own distributed systems both within and between sites, and telecommunications and information technology have become essential, business-enabling technologies so that the success or failure of an enterprise can depend on the availability and the reliable provision of fast and cost-efficient global communications. In many cases firms are dependent on networking capabilities to exist and survive [HPNRG]. In an increasingly competitive environment this trend is likely to continue. In particular, as networks become more essential to business, the possibility of outsourcing the corporate telecommunications management has become more attractive in order to allow companies to concentrate on their core business.

Businesses are themselves distributed over various sites and so require networking capabilities and advanced teleservices to provide connection and cooperation facilities between employees. In order to retain their competitiveness, organisations are promoting new working styles which require increased mobility from their employees or contractors: physical and/or organisational relocation, mobile working and telecommuting, which implies access to the corporate network from general public or residential network accesses. These new working styles require appropriate teleservices dependent upon a reliable and high quality telecommunications infrastructure.

1.1.3   An Emerging Multi-Stakeholder Environment

In the past, with regulated national telecommunications markets in which communication mainly occurred locally or within the same country, most aspects of the provision and management of a telecommunications service were under the full and exclusive control of the respective national telecommunications operator. Since the telecommunications operators usually had a monopoly on all telecommunications services this led to a situation where, at least as far as national communications were concerned, authority over all physical and logical resources was vested in a single telecommunications operator or its organisational units. The telecommunications operators were the exclusive providers for the existing telecommunications services and they were fully responsible for all the individual services within their servicing area. Therefore, there was little or no need to provide other organisations with access to the available proprietary management solutions. The few existing multi-domain management requirements resulting from communications across domain boundaries, such as the need for global billing, reverse charging, or address and telephone number inquiries, were usually resolved manually. These interactions could be characterised as horizontal or peer since they were limited to interactions between telecommunications companies with equal status. Since there was no natural technical distinction between bearer and value-added services, the need for a vertical or hierarchical structuring of domains did not exist.

In such a regulated environment the monopoly determined what services and level of service the subscriber received and this monopoly/subscriber relationship was implicit in how networks and services were designed. The advent of integrated broadband technology, the liberalisation of the markets, and the growing demand for global communications has changed all this and led to a far more complex domain situation where effective multi-domain management concepts and services are required. In a liberalised competitive environment with an increase in the number of services and service providers, services may well run over the networks belonging to several network operators. Service providers and network operators will need to cooperate with each other and with the end customer's management system in order to deliver and manage services on an end-to-end basis. The various stakeholders will be increasingly dependent on each other but will come from different backgrounds, experiences, domains, and perspectives. It is clear that the interactions and responsibilities between the various players must be clearly defined to prevent conflicts and mismanagement from occurring.

In considering TMN-based multi-domain management, the following organisational stakeholders are involved in providing, supporting, and using telecommunications services:

- The *network operators* running the networks and providing the basic communications capabilities (bearer services).

- The *service providers* offering additional (value-added) services over the basic communications facilities.

- The *customers* purchasing the services offered and who are either end customers who make these services available for use by end users, both business and residential, or add value and in turn offer the services to other customers.

At the organisational level, each organisation involved has particular responsibilities towards other organisations. The relationships between the organisations must be clearly defined via organisational boundaries over which certain formal procedures can take place. An enterprise model is therefore required which defines how each of these stakeholders can cooperate in offering, managing and using services. The application of domains to this model allows a structuring of the environment into clearly delineated areas with well-defined interfaces and enables the boundaries of autonomous participants to be determined.

1.1.3.1 Network Operators

Network operators are organisations operating various types of communications paths in the form of networks for public use. The issue of network operators owning networks distinguishes them from other service providers as they provide the basic infrastructure over which all other value-added services are offered. Public network operators are legally obliged to provide some specific services to the general public on conditions defined by regulation and legislation, i.e., the traditional monopoly telecommunications operators to which the ONA and ONP regulations apply, whereas private network operators are not subject to such provisioning obligations. These may be owned by companies who wish to sell network capacity superfluous to their own requirements, such as cable television or utility companies. There is one bearer service provider domain per network operator which makes available basic network services running on different types of network technologies. At least one network operator domain must be involved in the end-to-end provision of a service to a customer. Cooperation between network operators usually takes place on a peer-to-peer basis and network services are offered to all service provider domains equally. Even though the network operators own networks, their main role as seen by the customers is as access providers because they provide access to regional, national, and international communications networks. Each network operator utilises different means to provide the access in terms of access networks.

1.1.3.2 Service Providers

The service providers are organisations that add specific value to the bearer services provided by the network operators. Service providers offer a variety of services, including:

- Telecommunications services, such as virtual private network (VPN) services.
- Information services.
- End user services, for example, multimedia teleservices and entertainment services.

The service providers rely on the bearer services provided by the network operators so that the relationship between the network operators and the service providers is that of service providers being customers and network operators being providers (of bearer services). Each service provider may interact with several network operators, or with one network operator offering one-stop shopping facilities.

The need to provide services that customers want and will actually pay for will increase competition and lead to both service providers and network operators having to introduce new services more rapidly, more efficiently, and more flexibly while at the same time reducing costs and improving service quality [Willetts]. They must be able to customise and manage the services they offer on an individual basis and provide customers with management functionality enabling them to monitor, control, and optimise the services they pay for. This requires service providers and network operators to introduce more sophisticated and automated procedures into their management of services in order to meet such customer demands.

1.1.3.3   Customers

> "If a customer can't use it, it might as well be broken - it might as well not exist!" Arno Penzias [BernYuh]

Customers are organisations subscribing to and paying for services which are either used by users belonging to their customer domain or which comprise part of a service that the customer offers as a service provider to other customers. They might own customer premises networks which they manage. A customer domain is defined in the context of the particular service subscribed to and may consist of one or more (globally distributed) sites. A service provider may also provide services to cooperating autonomous customer domains, for example, various organisations cooperating on a particular project. Customers require a single point of contact for external service provisioning which is provided by the service provider domain.

Technological changes, liberalisation, and business trends have resulted in an increasingly significant role for customers and users. Customers, and particularly corporate customers for whom responsive and efficient telecommunications capabilities are essential for their core business, are becoming more demanding and knowledgeable concerning the services they purchase. They are no longer prepared to accept whatever the local monopoly offers and are able to pick and choose from a global market. They must see the benefit from subscribing to a new service or to new features in an existing service, and this must be at a price that they are prepared to pay. Services are judged not only according to cost, but also according to their quality of service, which is defined as user satisfaction with the service as it is perceived at the user interface [Seitz]. Customers expect high levels of connectivity, bandwidth on demand, convenience, dynamic response, and services tailored to their specific requirements, and they will select the services that most closely meet their requirements. There will be real-time requirements and a demand for reliability and robustness as customers depend on these services to carry out their core business. Customers are therefore playing an important role in driving trends in the market.

Customers are not only making demands concerning delivery and use of services but also on the facilities available for monitoring and controlling the services they purchase. In a competitive environment, customer requirements concerning the control that they have over the services they purchase may well influence purchasing decisions. A basic requirement is therefore access to information on the status, performance, fault, and accounting statistics of these services. They also want more active control over their services, including access to management functionality and customer control features, such as changing service parameters to ensure the service delivery that best meets their individual needs. They may wish to be able to change their configuration easily and quickly, such as increasing bandwidth on a link in order to prevent a potential congestion; and in order to keep costs within acceptable bounds they may want access to usage and charging data and to be able to determine low-cost routes. Service provider management systems must be able to provide an automated, secure interface accessible from customer management systems and supporting the functionality required in order to rapidly and flexibly carry out customer operations.

1.1.3.4   Stakeholder Relationships

The customer/provider roles associated with each stakeholder determine the relationships between the organisational domains. The role of each stakeholder is defined by its rights, responsibilities, and obligations with respect to other stakeholders and to the services and resources being managed. There can be a hierarchy of services and of customer/provider relations as relationships between the domains are established. The users can expect to have uniform access to services no matter where they are or what the underlying network provider domains are. However, this customer/provider relationship occurs at several levels; there are the providers of the networks and the customers of the network services. Customers can add value to the bearer services by providing value-added services, such as VPN services. Thus, a customer of network services can act as a provider of other services to different customers. Services from one provider are used by other providers as components in their services, resulting in longer value-added chains of services (see Figure 1.2). This customer/provider hierarchy can be extended as, for example, a customer of a VPN may offer several different services and these services may be further used to deliver a final service to end users. To enable the provision of stable and open services with this customer/provider model, access to management systems at each level in this hierarchy is vital.

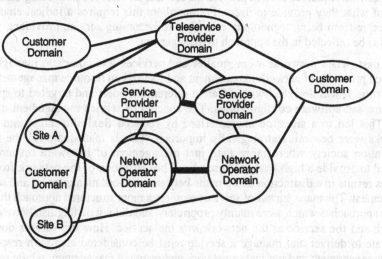

——— Hierarchical relationship between service provider and customer.

▬▬▬ Peer-to-peer relationship between providers of similar services in different administrative domains.

Figure 1.2: Relationships between network operator, service provider, and customer

In addition to these vertical relationships, the fact that service providers may not offer complete global coverage on their own and may need to enter into cooperative agreements in order to deliver a service to users means that horizontal relationships exist between the various providers of similar services (see Figure 1.2). These are peer-to-peer relationships as the service providers are cooperating to ensure the end-to-end delivery of any one service. These horizontal relationships also require that the cooperating providers provide each other with access to their management systems.

So although the environment is liberalised and competitive, the cooperative relationships between service providers, network operators, and customers are vital when providing end-to-end services, especially in order to react flexibly to the customer's requirements for tailored service delivery. The introduction of advanced services should not be hindered by the lack of coordination between the stakeholders regarding management of these services. There must therefore be a consistent management understanding and functionality across all domains involved in managing such services.

1.1.4   The Focus on Services and Service Management

> "The telecommunications industry, which has been interconnection-driven, will, in the future, be service-driven!" [BenPol]

The global information society is an open service market driven by customer and user requirements. This implies that customer and user perspectives should influence how the functionality of individual services and their management is designed. Instead of the traditional bottom-up technological approach where a service was defined in terms of the network resources used to provide the service, services are now being defined in terms of what they provide to users. For suppliers this requires a radical change of perspective from being technology providers to becoming service providers, which must also be reflected in the approach to management.

In the past, when networks were smaller and services less complex, management consisted primarily of network management and focused on resource management and monitoring. It was designed with different reference models and targeted to specific hardware and software configurations. The concept of service management did not exist. This led to a situation characterised by reduced flexibility at a time when networks were becoming strategically important. This is inadequate for the future information society where more than just management of hardware resources is required to provide a high quality teleservice. The separation of the network from the services results in a distinction being made between network management and service management. The management of services requires a more structured approach than the former approaches which were mainly proprietary and did not distinguish between the network and the service as the network *was* the service. How the various domains cooperate to deliver and manage a service must be considered also with respect to service management and not just to network and resource management where most of the work to date has concentrated.

This means that service management is becoming more important as the service market expands. Service management is concerned with ensuring that services are provided to customers efficiently and at the quality of service required and also with administering the customer subscription details associated with the provision of the service to the customer, such as customer information, accounting information, billing procedures, quality of service agreements, etc. Service management is more demanding as it is not only visible to customers but it must integrate, filter, and build upon the management of the individual resources, networks, and other services comprising the infrastructure supporting the service. Each of these components can have an effect on the management of the service, as can the performance of individual components in combination with others. Information on performance and faults must be correlated in order to understand their effect on the service performance, usage of the resources must be correlated with individual customers, sufficient transmission capacity must be available for the service to run, etc. In order to be able to determine the performance of such components, a high-level view is required that can relate the lower layers to that required by the customer, and which is sufficiently technology-independent to be applied across many different technologies so that the management of components in every domain can be transmitted to the service layer in a format that will be understood at this level [NMF-SF].

1.1.5 New Challenges for Management

The characterisation of the global service environment in the previous sections has shown that far-reaching changes are taking place in the telecommunications world. The infrastructure for managing a diversity of interconnected networks and for creating and managing the services that run over them has to offer more powerful management support than that needed for the traditional single-service networks and is not in place today. It has been estimated that the introduction of multimedia into the networks increases complexity tenfold, and a thirty fold complexity is on the horizon [BernYuh]. Advanced multimedia services have stringent requirements concerning bandwidth and latency and often require a guaranteed level of service reliability, especially for video and audio based services. Services will be expected to be open for individual adjustments, possibly performed in real-time, and maintained by the provider of that service. Real-time user control of these services requires new and powerful solutions. They need to be managed on an end-to-end basis and require an advanced, highly automated management infrastructure that can support their requirements.

The globalisation of the environment requires the participation of several domains in delivering a service to users. Each domain can belong to a different organisation, consists of equipment provided by a variety of vendors and supplying a variety of services, and has its own autonomous management system. But customers and users want real-time performance and quality of service from the service provider regardless of these underlying domains and management interoperability problems. Seamless provision of services can be achieved only if the underlying infrastructure supports the necessary management functionality to ensure that the networks and services and the individual domains can interoperate to deliver the service at quality of service levels satisfactory to both customers and users. In addition, service providers need to meet customer requirements by enabling customer management systems to carry out query and control operations on the services subscribed to.

Traditional network management approaches cannot cope with the explosion in the number and type of service. The technologies and standards for management of the services and infrastructure as one integrated whole are still immature and lacking essential features to ensure convenient interoperability and the flexible management required for the advanced multimedia real-time services being used. Management of individual resources does not provide adequate support for managing these multimedia services on an end-to-end basis. The variety of existing management concepts, each with its own management functionality and information models, has resulted in incompatibility between the different types of management system and restricts their interoperability. Advanced service management will not be possible unless the management systems in the various domains can interoperate with those of other domains and management services can interoperate with other management services. Indeed, it is only with integrated, automated end-to-end management capabilities that it will even be possible to realise the global information society of the twenty-first century. New concepts in managing networks and services must be developed that correspond to the demands of the future, requiring a paradigm shift that takes into consideration the fundamentally different characteristics of the new environment and how services will be provided and managed in it [Ejiri].

The provision of an effective multi-domain management infrastructure based on a suitable architecture is a necessary prerequisite for supporting advanced services in the global information society. The need for multi-domain management is a logical consequence of introducing integrated communications networks into a liberalised and highly competitive market environment. By investigating the problems of multi-domain management the work reported in this book is aiming to contribute to our understanding of how to manage the global information society of the next century.

1.2   TMN-Based End-to-End Service Management

In the course of providing or consuming telecommunications services, all stakeholders have certain management requirements. Customers require some influence over the services they purchase in order to ensure a certain quality of service or to query or modify service-relevant parameters, for example. Service providers on the other hand need to keep track of service usage and have to protect themselves from unauthorised service consumption. They will need to utilise advanced networking and management systems technologies in order to remain competitive. Network operators will need to deploy management systems that enable them to provide customers with a high quality of service and an end-to-end throughput, matching or exceeding that which is offered by current high speed LAN technology.

An end-to-end service is understood to mean reaching from the source(s) of delivery to the end user. End-to-end service management therefore involves network service providers and possibly other service providers as well as the end customer organisation and the service provider interfacing to the end customer. Each organisation has its own intra-domain, or internal, management system but this accumulation of intra-domain management systems is not sufficient to dynamically manage the resources from different providers for end-to-end service provision. If services are to be provided to customers effectively and with the quality of service requested certain management functionality is required, for example, on-line administration and configuration of services, rapid problem detection, automatic repair, or trouble ticketing, performance monitoring and optimisation, accounting, and security. Advanced on-line management capabilities are required to adapt to dynamically changing quality of service requirements of sophisticated multimedia teleservices.

End-to-end service management is an extremely challenging task. There is a considerable set of problems to be solved that are concerned with, for example:

- Designing a management architecture and appropriate algorithms for end-to-end service management.

- Identifying domains and functional components communicating over domain boundaries.

- Defining management interfaces for external access.

- Modelling management information at different provider levels and establishing shared management knowledge.

- Achieving security by employing authentication and access control mechanisms.

The most relevant issues that have to be considered and resolved before end-to-end service management can be realised in any sensible and meaningful way are discussed below.

1.2.1 Standardised Architectural Framework

As the demand for more advanced services grows, a standardised approach to the way networks and services are managed is required by telecommunications service providers and customers alike. Cooperation between customers, service providers, and network operators to achieve end-to-end management of services across several autonomous domains implies that a large number of actors depend on access to management functionality and services. An open architecture with standardised interfaces is essential so that cooperation can take place on a clearly-defined basis. Standardisation of the interoperable interfaces provides the facility to integrate the islands of proprietary management systems within both single-site organisations as well as in national, continental, and global enterprises.

A modelling approach is therefore required that can:

- Structure the management functionality in an architectural framework so that the control and management functions can be distributed over the many domains involved.

- Designate the management services available at each interface in the architecture.

- Specify the management information model that is needed in each domain.

The definition of a suitable architecture to facilitate end-to-end service management with multiple providers involved has to take various aspects into account, for example the business structure (reflected in the number of enterprises defining autonomous domains), the model of end-to-end communication functionality, and the model of functionality implemented or represented within the individual domains.

In the work described in this book the TMN recommendations are used as a framework for integrating network and service management activities in both private and public domains (if necessary using mediation and/or Q adaptor functionality)[1]. The use of TMN principles enables the relationships between the basic operations system function blocks to be defined via standard, interoperable interfaces. Multi-domain

---

[1]   See Appendix A for an overview of the TMN recommendations.

management is modelled in terms of interworking operations systems (OSs) and the TMN architecture is defined so that there is a clear distinction between inter-domain management and intra-domain management communication. According to the TMN recommendations, the interface where TMNs from different domains interwork is the X interface, and the growing demand for interoperability between the various service providers and network operators means that this interface and the way in which cooperative management is carried out over this interface is becoming increasingly important. Management communication between domains occurs over interoperable TMN X interfaces which can exist between each physical instance of an operations system function (OSF) at a particular management layer within a single organisation and its peer instance within partner organisations and customer organisations. Precise and widely accepted specifications of TMN X interfaces are required to improve the monitoring and control capabilities within global communications infrastructures and to achieve the goal of end-to-end communications.

1.2.2   Interoperability

"Two features are essential to the deployment of the information infrastructure needed by the information society: one is a seamless interconnection of networks and the other that the services and applications which build on them should be able to work together (interoperability)." [Bangemann]

Interoperability is a key issue in managing services spanning several domains, and is particularly important where no single provider is able to satisfy all the customer's needs, especially with respect to geographical coverage. The end-to-end management of services depends upon the exchange of management information and the provision of advanced management capabilities between several stakeholders, including the customers themselves. Service providers need standardised interoperable interface specifications in order to provision and manage services on a global basis in a heterogeneous environment. As service providers wish to market their products to as many customers as possible it is clear that they have a strong business interest in achieving interoperability [Vecchi].

Interoperability is very difficult to achieve because of the different protocols, and different versions of the same protocols, that are in use and different management systems in each domain which hinders a consistent and common understanding of the management functionality required [Albanese]. Interoperability involves more than interoperable communications protocols. If management services are to interoperate between domains, each management system from every domain involved needs to have the same understanding of the semantics of these services and of the context in which they are used. This means that the inter-domain management interfaces must focus on inter-domain interoperability issues (see Figure 1.3). Each inter-domain boundary represents a significant point where interoperability problems have to be resolved, i.e., the boundary between service provider and customer management systems, the boundary between network operator and service provider management systems, and the boundary between peer-to-peer management systems.

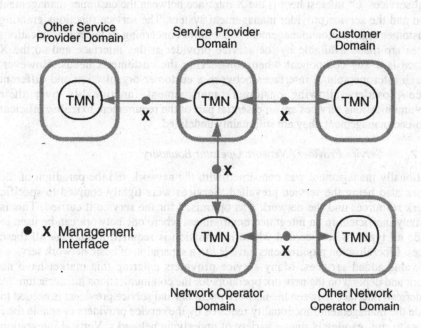

Figure 1.3: Service providers, customers, network operators interoperating via management interfaces

1.2.2.1   Customer / Service Provider Boundary

With end-to-end service management reaching from source to end user, management cooperation must take place between the providers and their customers. Corporate customers have established increasingly sophisticated management systems for their internal, private networks that provide a similar functionality to that available to service provider and network operator management systems, and service providers can offer management services that take advantage of this sophistication. Some management capabilities may also be offered directly or indirectly to users. The nature of the management cooperation will typically be customers making use of provided management services, but in the case of outsourcing providers may be operating the customer networks themselves.

Corporate customers are becoming more involved in telecommunications as their use of telecommunications capabilities increases and they have to be able to integrate their local management systems with those of the service providers and network operators if they wish to manage and control the services they use from their own management system. Therefore, the corporate data communications world, with its widespread use of the Simple Network Management Protocol [SNMP] and proprietary functionality over LANs, and the telecommunications world, with its focus on wide-area switched networks and the TMN recommendations, have to be integrated in order to support end-to-end management.

Customer management systems may interwork with provider management systems at one or more levels of service according to the services provided, the extent of one-stop shopping, and in general, the value added by one level's services relative to other levels' services. Of interest here is the X interface between the customer management system and the service provider management system. The service functions granting the customer organisation management capabilities concerning its telecommunications services are made available by the service provider at this interface and so the X interface is a key component when considering the customer's needs. However, although interoperable X interfaces between a customer organisation and different service providers allowing customers management functionality over their telecommunications services are an essential part of the requirement to realise efficient end-to-end management, they are still mainly undefined.

1.2.2.2   Service Provider / Network Operator Boundary

Traditionally management was concerned with the network and the paradigm of the network also being the service prevailed. Services were tightly coupled to specific network resources and the network was optimised for the service it carried. This is extremely inefficient in an integrated environment where one network can be used to provide all types of service and where information is required to manage all these services. Liberalisation requirements have led to a separation of basic network services from value-added services. Many service providers entering this market have no network and depend on the network operators for the communications infrastructure. A well-defined interface between the network operators and service providers is needed to provide the management functionality required by the service providers to enable their services to run seamlessly over a variety of underlying networks. Vertical integration of management services from the network element layer and the network layer[1] through the various intermediate services up to the management of the end user teleservices is required in order to guarantee end-to-end service quality levels, which implies that the OSs in different management layers are able to interact on the basis of the same management information [Okazaki].

Service monitoring at the network operator level includes monitoring network parameters and mapping to negotiated quality of service parameters at the service layer. This mapping has to be done if, for example, threshold values are exceeded and result in a notification (attribute value change) to the service provider's management system. The network operator management systems must therefore contain both service layer and network layer information pertaining to the services they offer. These management systems in turn control the network elements in their respective domains. The internal organisation of the network operator management systems is largely a decision of the network operators, although it is important to specify the information and functionality made available to service providers in line with open provisioning concepts. The service layer information model of a domain describes the information relevant to manage the basic communications services provided by the network. This is the model that is visible to the service provider across the X interface between the service provider management system and the network management system.

1.2.2.3   Peer-to-Peer Boundary

Peer-to-peer relationships occur where providers of a similar service cooperate in providing an end-to-end service to a customer. End-to-end management of the service

---

[1]   For information on the various management layers in the TMN Logical Layered Architecture see section A.3.4 of Appendix A.

is based on peer-to-peer cooperation between individual providers, whether network operators or providers of value-added services. This means that they need to offer an amount of cooperative management control to each other through open management interfaces. The relationship between any two providers can range from the customer-service provider relationship to a close alliance relationship. The management functionality available between any two peer providers will to a certain extent depend on the relationship between them, but at a minimum there must be available over the X interface between their management systems the management functionality required to allow the two domains to cooperate in managing any service that runs over both their domains.

1.2.3 Management Services

A standardised architectural framework is required in order to support the full range of end-to-end management services. The services provided at the X interface should between them be capable of offering a range of management functions as users of such services will have different requirements on the management services, depending on the types of service being managed. The abstractions developed for the service should be at a high enough level to offer a simple view of the service to human users and to be independent of any underlying network technology. As the management services need to interoperate between the various domains and require the support of these domains' management systems, it is essential that the services are clearly defined so that a common understanding of their semantics and syntax can be achieved by those offering the services and those using them. Some management services have started to be defined within the TMN framework, mainly for the management of telecommunications equipment [M.3200]. However, supporting the end-to-end management of services across domains is more complex, requiring an understanding of the end-to-end service being managed and the context in which the management services are being provided [NMF-BPM].

Many of the multi-domain management services will be concerned with the same systems management functional areas [X.700] as intra-domain management. Fault, configuration and performance management are essential to all management, whether intra-domain or multi-domain. However, the areas of accounting and security management are of particular importance in a multi-domain environment and require a higher level of support than is necessarily the case in an intra-domain management system.

Accounting and Billing. In a global service market with several cooperating providers at different service levels it is crucial to design and implement flexible accounting schemes and to record accounting information at the various levels. The service being charged for will often be quite different at the various provider levels and so the accounting information will vary, reflecting the variety of resources utilised at different service levels.

Accounting management deals with the financial side of service provisioning in relation to the customer. The financial conditions between provider and customer are defined in a contract which can include customer information, tariff details, billing information, and any agreed accounting limits. Also included can be maintenance arrangements, rental charges, security, quality of service requirements and guarantees, and rebate schemes in cases where service quality guarantees are not maintained. The customer is periodically invoiced by a bill with the amount due, including details of service usage, information on the balance of the account, credit rating and limits, etc. This procedure needs to be supported by the service provider's management system and can be based on the standardised *Usage Metering Function* [X.742] where the accounting process is broken up into three subprocesses: the usage metering process, the charging process, and the billing process. Management information relating to these processes must be modelled and agreed between the various domains involved in delivering a service.

Security Issues. Services crossing several networks and organisational domains will only be used if they can support customer security requirements concerning confidentiality and integrity of the data being handled both by the services themselves as well as by the management services offered to the customer. Security alarm reporting, audit trails, authentication, and authorisation are essential and guaranteed security levels must be offered if customers are to feel confident in using the services offered. Any information relevant to the customer and its use of a service, and this includes the management information associated with the service, must be confidential and made available only to authorised users who have been authenticated.

From its position within the TMN architecture it is clear that the X interface imposes stringent requirements on security. Even if the scope of management operations is considerably confined by the supported information model, the information conveyed over several protocols should not be visible for anybody from outside. Intruders should not be allowed to retrieve private information or perform actions that corrupt the service or the operation of the network. It is therefore necessary to ensure the availability, accountability, confidentiality, and correctness of the management of data accessible to other management systems over the X interface.

1.2.4   Management Information Modelling

The modelling of management information crystallises the multi-domain management problem as it is the management information model that supports the management services available by determining how resources can be managed and what management functionality is available to manage any one of these resources. It is the information models mutually supported by each of the communicating management systems that define the range of management operations inside each domain and enable the management services defined to interoperate over the X interface to be used meaningfully and consistently. In order to support management cooperation between the autonomous management systems a considerable amount of information needs to be available in the various domains and exchanged between them across the X interface. Moreover, this information must be interpreted by all participants in the same way and be able to support cooperative management interactions between the domains. If these requirements are not met interoperability between management systems is unlikely to take place.

In order to support one-stop shopping (end-to-end service provisioning), one-source billing (only one service provider issues the bill), and one-stop complaining (end-to-end performance and maintenance) management services require the exchange of configuration, fault, security, accounting, performance, and maintenance management data between the different domains involved. In addition, customers, particularly business customers, will require customer management services which provide access to the service and network management information of the services that they have subscribed to.

If end-to-end service management is to be achieved, each domain involved has to support the relevant parts of the end-to-end management concepts. Each domain's management system has to expose a view about its communications resources and services to the outside world. There must be some way of ascertaining what management information exists and how this information is modelled. As the TMN recommendations have adopted OSI Systems Management standards for management information modelling and exchange [M.3010], each information exchange deals with managed object (MO) manipulations or information retrieval from MOs. The communicating management systems must therefore share a common view of the management information and the management functions. This is referred to as *shared management knowledge* and includes the semantics of the information, the MO classes, instances, and functions supported, containment relationships and name bindings, as well as a common understanding of the scope of management operations and the relationships between the individual information entities [X.701]. By exposing management information and management operations at the X interface, restricted control is given to an external entity. The placing of this external access at the service layer is reasonable. It is then still up to the responsible management system within the relevant domain to map these (inter-domain) service requests appropriately onto the (intra-domain) network layer and the network element layer. An outside manager needs to know about the abstract services provided by the network and not about the details of network operation and technology.

These requirements can only be met if the underlying information models are adequate. This involves both shared management knowledge as well as a common understanding of the management architecture, the domains, and the management policies of the domains. The information must be accessible in the individual domains and it must be clear what management operations are available over the X interface for manipulating the information. An integrated information model which makes available a common view on all information types relevant to multi-domain management should provide human managers and management services with the information they need to perform their management tasks. The main areas where information additional to that contained in current management standards (OSI Systems Management, TMN, SNMP) is required are identification, location, and addressing. The management components need to be locatable and addressable wherever they are stored in the global environment and so a logical, integrated naming and addressing scheme for identifying and locating management components and information in the form of managed objects needs to be defined. A common name space must be constructed to allow global identification of resources by unambiguous names, and name and address resolution must be provided so that management information users can access information by name.

As well as an integrated information model that provides the functionality required for multi-domain management, it has to be decided where and how to locate the information as it is distributed over several domains and located in various end systems. The management of end-to-end services requires that management operations are invoked over all the relevant domains with information from several management information bases being accessed. End-to-end service management therefore requires a global information architecture which includes the following considerations:

- Accessibility to service information throughout the distributed environment.

- Identification and location of objects relevant to the provision of service information, i.e., the unique and global naming of these objects.

- Realisation of a common name space.
- Cooperative administration of this common name space.
- The operations to be invoked and the results to be received.

The dynamics of the environment imply that relationships can be established between domains that do not have any specific knowledge about each other. The information needed for end-to-end service management has to be available to all cooperating domains and so it must be maintained and made accessible by a globally available information service which must therefore be based on open standards. The TMN framework needs to provide support for such an information service for storing, accessing, and maintaining globally relevant information on, for example, service providers, their offered capabilities, contact names and addresses, and operational information, such as communication addresses of OSs, information models, and other information related to shared management knowledge.

1.3 Summary

This chapter has introduced the characteristics of the global information society and outlined the issues concerned with managing that environment. Technological changes leading to high speed networks and the ubiquitous use of advanced services coupled with liberalisation of the telecommunications market and increased business dependence on communications is creating an environment in which effective multi-domain management concepts and services are essential. This is providing a challenge to traditional management approaches which were developed primarily for single domains where the service was tightly coupled to the network it was running over. The introduction of many of the issues that must be considered when developing management systems for the global information society has shown that multi-domain management is not a trivial task and that it can only be achieved if all stakeholders support the same understanding of the management functionality operating over the interfaces between their management systems.

The rest of this book is concerned with the specific work undertaken by the PREPARE project on TMN-based multi-domain management, its experiences, and the results it achieved.

2 The PREPARE Project

2.1 Introduction

Of fundamental importance to the results presented in this book is the overall approach of the work in PREPARE which was centred around the verification of results through the implementation and testing of theoretical principles. This chapter introduces the PREPARE project.

First the project is described within the context of the RACE (Research and development in Advanced Communications technologies in Europe) programme (section 2.2), then the project consortium is presented (section 2.3). In section 2.4 the main objectives of the project are summarised, and section 2.5 outlines the overall approach taken by the project. The starting points and working assumptions are described in section 2.6. Sections 2.7 and 2.8 give overviews of the two main project phases. The two project phases are characterised by the *organisational situation* considered, the *broadband network testbed* being managed, and the *multi-domain management system architectures* designed for testing purposes. In addition to describing the context in which the results are presented, these aspects of our work constitute a common basis for the examples provided in the remaining chapters of the book. Finally, section 2.9 provides a summary of this chapter.

2.2 PREPARE and the RACE II Programme

The PREPARE (PRE-Pilot in Advanced REsource management) project was set up as part of the RACE programme of the European Union (EU) to investigate network and service management issues in the multiple bearer and value-added service provider context of an open telecommunications service market.

The RACE programme was established, and partly sponsored, by the EU with the objective to "promote the competitiveness of the Community's telecommunications industry, operators and service providers in order to make available to final users the services which will sustain the competitiveness of the European economy and contribute to maintaining and creating employment in Europe." RACE work is focused around developing aspects of *Integrated Broadband Communications*[1] (IBC) [RACE].

The work in RACE is characterised as "pre-competitive and pre-normative R&D cooperation typically between organisations established in different Member States (horizontal collaboration) and between technologists, operators and users (vertical collaboration)" [RACE]. This implies that projects in the RACE programme are based on collaboration between various types of organisations, each playing one or more defined roles in the project, and all working towards a set of common project goals. These goals are ultimately defined as part of the overall RACE programme goals, and each project employs its individual approach to reaching these goals [RACE].

---

[1] "IBC is a global concept that covers all kinds of communications and technical and operational means to offer services." [RACE]. In this chapter, IBC network means a telecommunications network supporting IBC services.

2.3   The PREPARE Consortium

The PREPARE Consortium brought together eight organisations having complementary expertise which would be necessary for realising real life inter-domain management systems. The project partners and their respective roles were:

- A network operator (Tele Danmark A/S) interested in integrating wide area network management with multi-domain service management based on Telecommunications Management Network (TMN) principles.

- A network equipment vendor (DSC A/S) interested in the management of ATM network elements and networks and the management of heterogeneous network interworking.

- A customer premises network, transport network, and management platform vendor (IBM) interested in applying and enhancing its products in a multi-domain environment.

- A vendor of network management platforms (L.M. Ericsson A/S in cooperation with Broadcom Eireann Research) interested in the application of their management platform to value-added service provision.

- Researchers into advanced network management techniques (University College London, Marben, and GMD-Fokus) interested in applying their platforms to the multi-domain environment.

- Researchers into multimedia applications (University College London and GMD-Fokus) interested in the interactions of these applications with service and network management.

Each project partner brought its own specific interests to the project, sometimes overlapping but often different or even contradictory. Therefore, although we were not operating in a true commercial environment, the viewpoints of the customer, the value-added service provider, the public network operator, the management platform vendor, and vendors of network technologies were all represented.

2.4   Objectives of the PREPARE Project

Management cooperation between organisations in the telecommunications service market is a prerequisite for an open service market. The objectives of PREPARE were addressing central aspects of such management cooperation. These were:

- *Definition* of a multi-domain management architecture required to support the creation and control of end-to-end services, such as virtual private network (VPN). Based on the TMN principles (which are described in Appendix A) this architecture in particular addresses the specification of interfaces in terms of communications protocols and managed objects.

- *Specification and implementation* of a *communications management demonstrator* providing a realistic environment for exercising services and interfaces, thus verifying the achievements of the architecture definition.

- *Standardisation* of the inter-domain management interfaces, seen as PREPARE's contribution to the broader RACE goal of contributing to standards.

The impact of the PREPARE work was to be the availability of practical and tested techniques that could be directly applied to the end-to-end management of services in the European environment of multiple organisational domains with a variety of network technologies. The availability of these results was expected to have a considerable influence on the success of the IBC in Europe.

2.5 Approach of PREPARE

A particular feature of the PREPARE project was the pragmatic and implementation-oriented approach aiming at fast verification of theoretical assumptions in the broadly defined area of standards-based multi-domain management.

This implied a natural and common-sense development life-cycle through the identification and analysis of multi-domain management requirements, the design of theoretical solutions and the subsequent implementation of these solutions in a communications management demonstrator based on a broadband network testbed. In this way it was possible to verify the solutions by test runs and public demonstrations of the demonstrator.

As a four-year project we had the opportunity to split the work into two phases. The first concentrated on fundamental aspects of multi-domain management such as the architecture, communications techniques, description techniques, the technologies employed for telecommunications management, and the heterogeneity of such management system technologies. The second phase concentrated more on applying the defined principles in a complex but also more realistic communications management demonstrator, which enabled more aspects to be shown both of the services relying on multi-domain management and of the complexity of the telecommunications service market.

One of the most important characteristics of PREPARE was the practical validation of the theoretical solutions. For this purpose, two broadband network testbeds were defined, one for each of the two project phases (see sections 2.7 and 2.8). The testbeds were based as far as possible on existing hardware and software so allowing us to concentrate on the core objectives rather than on constructing new network components, management platforms, etc.

2.6 Starting Points and Working Assumptions

2.6.1 International Standards Background

The primary goal of PREPARE was to develop and demonstrate a multi-domain management architecture that spanned several management administrative domains and was based on the TMN recommendations. The work was therefore oriented towards the use of open systems architecture and interfaces as defined by standards bodies such as ISO, ETSI, and ITU.

Many specifications based on the TMN framework have been produced by standards bodies. The majority of these specifications are interface specifications developed in

accordance with the *OSI Systems Management* standards[1]. They include generic
network management information models, such as the ITU recommendation *Generic
Network Information Model* [M.3100], as well as technology-specific information
models, such as the ITU recommendation on managed objects for signalling system
no. 7, *Network Element Management Information Model for the Message Transfer
Part* [Q.751]. However, when PREPARE started there were no generic service layer
models available. Likewise, very little existed for the specific technologies present in
PREPARE's broadband network testbed. Therefore, a large amount of work was
devoted to the definition of such information models.

The recommendations in the ITU's TMN architecture and OSI systems management
were however found not to provide sufficient support for some inter-domain
management activities and technical needs: a globally available information service
and a common naming schema for management information objects. To solve these
problems PREPARE developed a concept based on the X.500 directory standard
[X.500] which is described in Appendix C[2].

Recalling the project objectives we can say that the TMN principles define the TMN
structure whereas PREPARE worked on adding content (function) to this structure.
The architectures developed by PREPARE can be seen as applications of these TMN
principles in the particular context of multi-domain management of specific services
running over broadband networks.

2.6.2   General Organisational Situation

The complexity of the organisational situation addressed is illustrated in Figure 2.1.
This shows a number of different organisations and some of their possible commercial
relationships (i.e., contractual relationships) for exchanging telecommunications
services (see also section 1.1.3) and reflects the open telecommunications service
market being promoted by various forces.

Figure 2.1 provides the scope for PREPARE's work in the sense of identifying the
roles of organisations considered by PREPARE as well as identifying the possible
interorganisational relationships between them. The organisational roles are:

- Customers and end users.

- Value-added service providers (VASPs).

- (Public) network operators (PuNOs).

The choice of these organisational roles is intended to reflect the organisations'
respective roles in the service provisioning chain. The customer purchases a particular
service, the end user uses the service. A VASP is an organisation which provides
value-added services to customers and end users. Such services are created by utilising
other services and adding a particular value to them. A public network operator
(PuNO) is an organisation owning and operating a network and providing bearer
services based on such a network.

---

1   A brief overview of OSI Systems Management is given in Appendix B.

2   This concept was subsequently introduced into TMN and OSI Systems Management. See Appendix
      A, section A.1.4.

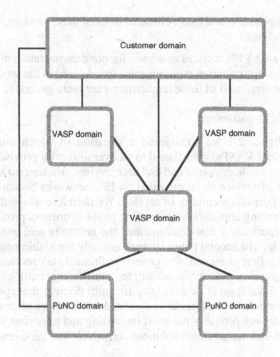

Figure 2.1: General enterprise situation

Two of the identified relationships between the organisations depicted in Figure 2.1, the relationship between a customer and a PuNO, and the relationship between two customer organisations, were not addressed in detail by PREPARE. In the first relationship it was considered that the effect of the organisations' characteristics on the possible management interfaces between the organisations' respective management systems was similar or even identical to those between a VASP and a PuNO. The second type of relationship was considered out of scope for PREPARE, although we have no doubts about its importance.

The two-phase approach allowed us to concentrate more narrowly on some aspects in turn. In the first phase we concentrated on the relationships between VASPs and customers and between a VASP and its providers (PuNOs) of bearer services. In the second phase we concentrated on the relationships between VASPs, between PuNOs, and between customers and multiple VASPs. Further details on the organisational situations addressed in the two phases are provided in the following sections.

2.6.3 General Customer Requirements

It was found very useful in our approach to try and develop some specific customer requirements since it was obvious during the work of PREPARE that even the very fundamental management architecture (cf. sections 2.7.3 and 2.8.3) was strongly influenced by such (mainly non-functional) requirements.

Non-functional requirements are requirements that do not relate explicitly to the type of service provided or to the possible usage or application of it by end users. Service requirements, on the other hand, relate to the nature and capabilities of the service being considered. This means that the main functional requirements are identified by investigating the application and usage characteristics of these services, including their management capabilities.

We defined IBC-based VPN services as a basis for our demonstration of multi-domain management and various customer requirements were derived in the specific context of VPN services. However, most of these requirements are more general in nature.

2.6.3.1   Service Requirements

As depicted in Figure 2.1 we considered a situation in which multiple service providers (PuNOs and VASPs) contributed to an overall service provisioning chain in order to create and provide customer and end user services. *Multimedia teleservices* are an important class of service to be provided on IBC networks because they require high capacity and guaranteed quality of service. We therefore adopted a VPN-based multimedia conferencing application for the first phase in order to provide the project with an end user application that demonstrated the necessity and benefits of multi-domain management. The second phase covered not only the multimedia conferencing application from the first phase but also general multimedia teleservice management. By looking at several services, for example, multimedia mail and multimedia conferencing, the development of concepts for multi-domain management of such services could be performed in a broader context and thus be more generally applicable. This class of service was probably the most interesting and important to apply and to demonstrate end-to-end management solutions for, which was the overall objective of PREPARE.

2.6.3.2   Non-Functional Requirements

- *End-to-end services.* PREPARE considered VPNs as an integral part of the business customers' information infrastructure (i.e., the full range of systems, equipment, telecommunications services, information servers, etc.). This implies that the VPN must extend into the customer premises if the customer so requires.

- *Outsourcing.* There is a general trend for companies to move away from being their own network operators and to focus their attention on their core business. Simplifying the management of a company's information infrastructure can be achieved by decreasing the amount of day-to-day operational work. This can be accomplished through outsourcing[1] parts, or all, of the information infrastructure. Outsourcing may be accompanied by management services for the customer which must provide fast, reliable on-demand access to management information and functionality, such as accounting information and configuration management services.

- *One-stop shopping.* Replacing internal operations with several external provider organisations is clearly undesirable since it would increase the complexity of having to deal with multiple providers, especially for global VPNs where considerations include different time zones, cultural and legislative differences, etc. For these reasons customers require one-stop shopping facilities from the

---

[1]   The concept of outsourcing means handing over responsibility for service provisioning to a third party [Lloyd].

VPN provider (one single point of contact for ordering, billing, complaining, etc.).

- *Cooperation.* In cases where a single provider is not able to satisfy all the customer's needs, the service provider must to some extent use subproviders, i.e., other service providers and/or network operators. These subproviders must provide not only communications or network services but also associated management services in order to allow the service provider to manage the service from end-to-end.

- *Resource administration.* Since customers are not interested in paying for infinite telecommunications resources (e.g., bandwidth), availability requirements make it necessary to offer end-to-end resource booking capabilities with guarantees for transmission capacity at predetermined times.

2.6.4 General TMN System Architecture

An important assumption was made throughout the design work of PREPARE. This was that each of the organisations owned some sort of technological system which contributed to the overall end-to-end service construction and its management. Each such system was either a managed network (or service) or a management system and was located in one of the organisational domains. With our focus on TMN we assumed that all management systems were *TMN building blocks* (see Appendix A). ITU recommendation M.3010 points out that there is a *"conceptual relationship between an administration and a TMN"*. Here, administration refers to *"public and private (customer and third party) administrations and/or organisations that operate or use a TMN"*[1] [M.3010]. In PREPARE inter-domain management is modelled in terms of interworking TMNs, with management activity between domains occurring at TMN X interfaces. We accordingly assumed that every organisation had its own TMN, including the customers which owned customer premises networks (CPNs)[2].

Figure 2.2 shows the TMNs in each domain and depicts their internal structure. TMN operations systems (OS) taking direct part in inter-domain management activities (such as service layer OSs, S_OSs) are interconnected by TMN X interfaces, whereas TMN OSs which do not take direct part in inter-domain management (such as network layer OSs, N_OSs) are connected via TMN Q interfaces. Networks and their interconnection are also depicted. (Some of the TMNs in Figure 2.2 do not manage networks directly, but rather manage services through interactions with other TMNs. These other TMNs may or may not be co-located with networks in their respective domains.)

The emphasis of PREPARE was on management cooperation between service providers and between providers and their customers. The term cooperation is here meant to cover both unidirectional service provision by one organisation to another organisation (*hierarchical cooperation*), as well as bidirectional service exchange between two organisations (*peer cooperation*). The particular aspect of peer cooperation in our understanding is that the services exchanged between organisations

---

1   The terms *administration* and *organisation* are used interchangeably in this chapter.

2   Sections 3.5.1 and 3.6.1 provide a detailed description of the architectural principles developed by PREPARE.

on a peer-to-peer basis are basically the same. This is for instance the case when several European PuNOs interconnect and cooperatively manage their national ATM networks in order to provide pan-European ATM services to customers. We use the term hierarchical cooperation when one provider organisation utilises another organisation's service and either adds value to it, or uses it to add value to the service it itself provides. An example of the former is the VPN defined by PREPARE, which adds value to bearer services such as the ATM virtual path (VP) service. Another possibility is that an organisation simply uses the service directly in support of various applications, as in the case of customer organisations with service end users in their domains.

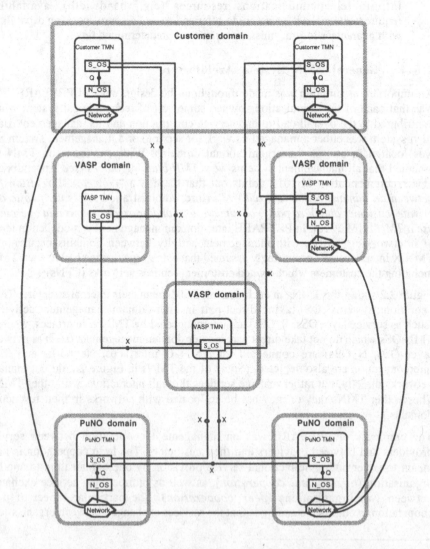

Figure 2.2: Illustration of PREPARE's TMN architectural approach

2.7   Overview of Phase 1

2.7.1   The Organisational Situation

The first phase of PREPARE considered an organisational situation involving four organisations:

- Two PuNOs providing bearer services over a metropolitan area network (MAN) and a wide area network (WAN).

- A VASP providing a VPN service by utilising and adding value to the bearer services provided by the two PuNOs.

- A customer organisation subscribing to the VPN service. End users are located in the customer premises and belong to the customer organisation.

The MAN and WAN networks are provided and operated by different PuNOs independently of each other from an organisational point of view. Moreover, the VASP offering the VPN service is organisationally independent of both the MAN and the WAN providers. This organisational situation is depicted in Figure 2.3.

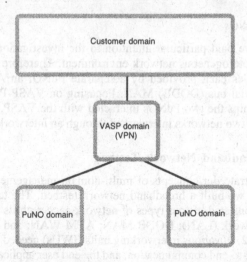

Figure 2.3: Organisational situation for phase 1

2.7.1.1   The Customer Organisation

The model of a multi-domain network used in PREPARE is that of CPNs interconnected by one or more public networks (PNs). Each of these networks is managed by a different organisation and this is reflected by each organisation possessing management operations systems at both the network and service layers.

The customer organisation represents a company using telecommunications services, such as a car manufacturer. It is distributed over several premises located in different European countries. The VASP provides the customer organisation with a VPN linking these premises. The employees of the customer organisation often meet via a multimedia conferencing application. These end users access the conferencing application at workstations located within the customer organisation's premises. The traffic generated by the multimedia conference application is routed across the VPN.

2.7.1.2   The VASP

The VPN service is provided not by the PuNOs but by an organisationally independent VASP to customer organisations which in turn own and/or coordinate more than one CPN. The VASP has its own TMN operations system which cooperates with the CPN service layer OSs to manage end-to-end services across the various PuNO domains and the customer domain utilised in the VPN service. Part of the value-added service provided by the VASP is to coordinate the management of the PuNO domains. This is performed on a contractual, financial, and administrative basis at the service management layer by having direct management links with the PN service OSs. The customer does not therefore have to implement any direct links, either organisationally or through a TMN, with the potentially large number of PuNOs involved in providing communications services between CPNs. An example of VPN management facilities that were made available to the customer is the management of the quality of service profile of communications resources available between terminal equipment on multiple remote sites (CPNs).

2.7.1.3   The PuNOs

In the first phase we paid particular attention to the investigation of multi-domain management in a heterogeneous network environment. Therefore we identified two different technologies each provided by a separate PuNO: an ATM WAN and a distributed queue dual bus (DQDB) MAN. Focusing on VASP-PuNO (and VASP-customer) relationships the two PuNOs interacted with the VASP, but not with each other. However, the two networks interworked through an interworking unit.

2.7.2   The Broadband Network Testbed

In order to demonstrate our concepts of multi-domain management in as realistic a setting as possible, we built a broadband network testbed. The testbed for the first phase (testbed 1) consisted of four types of network each with its own TMN: Token Ring local area network (LAN); DQDB MAN; ATM WAN; and ATM Multiplexer (MUX) (see Figure 2.4), various interworking units (IWUs) needed to interconnect the subnetworks for end-to-end communication, and the end user application.

The testbed carried two of the main traffic types of a future IBC network: connectionless traffic and isochronous connection-oriented traffic. For the connectionless traffic PREPARE based the implementation on the ETSI connectionless broadband data service (CBDS) standard [CBDS]. For the isochronous connection-oriented traffic, the standardisation of neither the ATM nor DQDB areas had reached a stable state. PREPARE therefore devised an approach to support the isochronous traffic where AAL1-like[1] connections extending across both the DQDB MAN and ATM WAN were used, and where connection set-up was done via management.

---

[1]   AAL-1 stands for ATM Adaptation Layer type 1, which is a protocol for constant bit rate transfer in an ATM network [I.362].

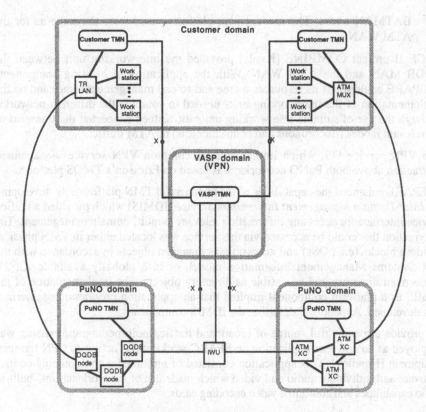

Figure 2.4: Testbed for phase 1

Testbed 1 used commercially available products and components originally developed within other projects. The networks and management platforms were as follows:

- *Token Ring LANs*. These were based on commercially available LAN products. They operated in the customer domain and were managed by IBM's NetView for AIX.

- *ATM WAN*. This was built for the Danish government project BATMAN [Nielsen] and consisted of three nodes at different Danish sites. These were first connected at 2Mbps, but were later upgraded to 155Mbps. The WAN connected Tele Danmark KTAS and the Technical University of Denmark in Copenhagen and Tele Danmark Jydsk Telefon in Aarhus. This network operated as part of the testbed's PuNO domain and was managed by OSIMIS [Pavlou] and ISODE [Onions] based systems at the network layer and Hewlett-Packard's OpenView at the service layer.

- *DQDB MAN*. This was a three-node network, located in Copenhagen and based on ETSI compliant nodes. It operated as part of the testbed's PuNO domain and was managed using OSIMIS and ISODE based systems at both the network layer and service layer.

- *ATM multiplexers*. These were based on ATM cross-connects from the BATMAN project and were intended to operate in a manner similar to ATM LANs. They operated in the customer domain and were based alongside

BATMAN nodes. The management platforms used were the same as for the ATM WAN.

RACE II project COMBINE [Kvols] provided the interworking unit between the DQDB MAN and the ATM WAN. With the application of existing components PREPARE could focus its resources on the end-to-end management issues and on the implementation of the interworking units needed to combine the different networks. Through the use of suitable interworking units the testbed supported the transport of the relevant modes (isochronous and connectionless) of ATM traffic.

The VPN service OS, which implemented a common VPN service management abstraction above both PuNO networks, was based on Ericsson's TMOS platform.

PREPARE enhanced the capabilities of the commercial TMN platforms by developing the *Inter-Domain Management Information Service* (IDMIS) which provided a unified service interface for accessing information relevant to multi-domain management. The information that could be accessed via this service was located either in TMN physical building blocks (e.g., OSs) and accessible as managed objects in accordance with the OSI Systems Management information model, or in a globally available X.500-conformant directory and accessible as directory objects. The implementation of the IDMIS as a platform component implied that an application programming interface was developed. Appendix C describes the IDMIS component.

To provide a meaningful source of broadband traffic, multimedia conferencing was employed as an end user application on SPARC workstations acting as CPN terminal equipment [Handley]. This application consisted of multimedia conferencing control software and individual audio and video which made use of the workstations' built-in audio capabilities and additional video encoding cards.

2.7.3   The TMN Multi-Domain Management System Architecture

The management architecture of the first phase closely followed the organisational situation outlined above. The VPN provider (i.e., the VASP) purchased services and interacted with corresponding management systems (S_OSs) in its subprovider domains. This was reflected in management interfaces between the VASP's S_OS and the two PuNOs' S_OSs (for ATM WAN management and DQDB MAN management). The management interactions of these systems followed the hierarchical cooperation mode, i.e., the PuNOs' S_OSs always acted in a managed role whereas the VASP's S_OS always acted in a managing role (see Figure 2.5).

In order to investigate the management of a truly integrated end-to-end service the VPN service management was not restricted to managing resources in the PuNO domains but also included customer network resources up to the terminal equipment as a result of cooperation between the customer domain and the VASP. This allowed not only the customer to access the VPN service but in particular the end users where appropriate.

The type of functionality required to manage end-to-end VPNs included the following:

- Specification, across any domain, of communication streams over which a certain quality of service is guaranteed for VPN customer traffic.

- Set-up and management of end-to-end communication paths spanning multiple domains. Although this type of functionality will be performed by signalling in true broadband networks, the only available option for global connection set-up in such a heterogeneous network was through coordinated multi-domain management.

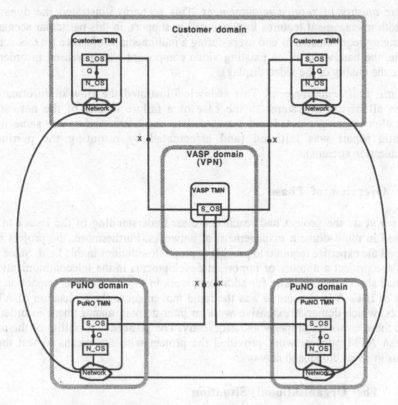

Figure 2.5: Management architecture for phase 1

To achieve this end-to-end management capability the VASP S_OS interacted with S_OSs in the customer domain. The mode of this interaction was peer cooperation in the sense that the VASP S_OS and the customer S_OSs could assume both roles of managed and managing system. For instance, a service request from one customer TMN to the VASP TMN would result in an associated service request from the VASP TMN to another customer TMN.

The TMN systems implementing the architecture shown in Figure 2.5 were shown at a live demonstration in December 1994. At the demonstration a number of multi-domain management scenarios were shown, for example:

Installation of a VPN. This scenario showed how a VPN management abstraction was created over all relevant testbed networks. Initiated by a VPN creation request from one of the CPN service layer OSs, the VPN service layer OS determined the construction of the VPN and interacted with the other CPN service layer OS as well as with the service layer OSs in the PuNO domains.

Multimedia conference set-up. This scenario illustrated the end users' access to management capabilities needed to establish end-to-end communication streams between participants in a multimedia conference.

Dynamic quality of service management. This scenario illustrated the dynamic bandwidth management features implemented to support, in this particular scenario, the changing requirements of end users during a multimedia conference (in this actual example, the bandwidth of an existing video component was increased in order to improve the quality of the video display).

Automatic fault management. This scenario illustrated the flow of information between all pertinent systems in the case of a failure in one of the networks. Notifications were generated and forwarded to end users, and at the same time automatic repair was initiated (and effectuated by rerouting the pertinent communication streams).

2.8   Overview of Phase 2

After two years the project had reached a clear understanding of the issues to be addressed in multi-domain management of networks. Furthermore, the project had developed the expertise required to provide appropriate solutions in this field. Since the start of the project a number of important developments in the telecommunications world had shown the necessity for additional work in this area and its impact on the success of IBC. A key example was the rapid movement to the realisation of ATM networks, which demands extensive work in inter-domain management in order to operate these networks properly and efficiently. The planned availability of the pan-European ATM pilot network provided the project with the means to test these solutions in a real broadband network.

2.8.1   The Organisational Situation

The organisational model upon which the second phase's management architecture and service scenarios were based consisted of customers with several CPNs which were interconnected by bearer services provided by multiple PuNOs. Over this configuration VASPs used the services offered by the public network providers to provide enhanced services to the customers. For the second testbed this model was more fully investigated with more complex interorganisational relationships being analysed and in some cases implemented as management services. The organisational situation for the second phase is depicted in Figure 2.6.

For the second phase we considered for the customer organisation a large shipping line with three sites, one in London, one in Berlin, and one in Copenhagen. At these sites the company had installed CPNs which were to be interconnected for various purposes. The situation which was considered in this context was the collaboration of ship engine experts located at the three sites to solve a particular problem with the engines of a ship at sea. Collaboration was achieved with the assistance of multimedia teleservices and applications, including multimedia conferencing, and joint viewing and editing of documents.

Several providers of telecommunications services were included in our overall framework:

Network operators. The network operators were providers of bearer services as well as value-added services (one-stop shopping, end-to-end across Europe). These network operators were traditional public network operators (PuNOs), one operating in the United Kingdom, one in Denmark, and another in Germany, that cooperated to provide a customer premises to customer premises ATM VP service with an associated set of management capabilities for the customer, and with both a single point of contact (for commercial and contractual interactions) and a single management system interface.

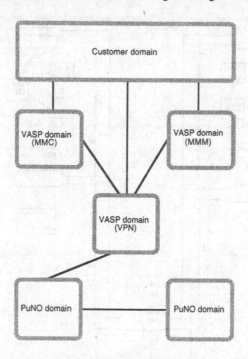

Figure 2.6: Organisational situation for phase 2

Independent providers of multimedia teleservices. Two teleservice providers offered multimedia conferencing and multimedia mail services to the customer organisation for use by end users.

Independent providers of value-added services. A VASP offered a pan-European VPN service based on its subscription to the above-mentioned ATM VP service.

The construction of value-added services of diverse types (for instance VPN, multimedia conferencing, and multimedia mail) can be performed in numerous ways by combining the characteristics of the services available. In particular, some value-added services may be constructed by adding value to other existing value-added services . In the second phase of PREPARE the two multimedia teleservices were offered to a customer which was already subscribing to the VPN service and which required the multimedia services to be delivered to VPN users. Thus, the multimedia teleservices were to be included into the overall portfolio of services available to end users. This resulted in complex relationships between the three VASPs, both with respect to contractual and financial off-line matters, as well as with respect to the exchange of on-line management information.

2.8.2   The Broadband Network Testbed

The sites included in the phase 2 broadband network testbed (testbed 2) were Tele
Danmark in Copenhagen, GMD-Fokus in Berlin, and University College London in
London. Interconnection between these three sites was achieved via the pan-European
ATM trial network. As opposed to the first phase, in which both ATM and DQDB
technologies were used, this second testbed was entirely ATM-based.

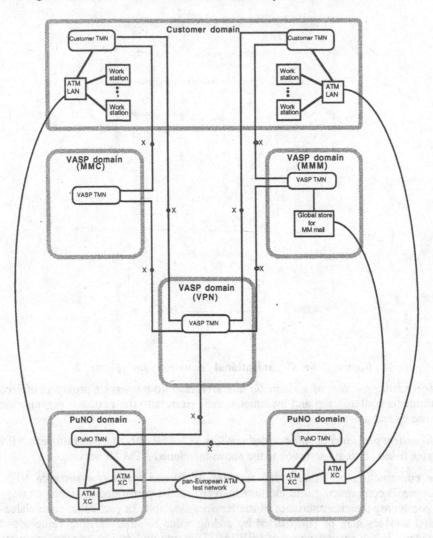

Figure 2.7: Testbed for phase 2

In order to continue reflecting the model of multiple CPNs interconnected by multiple
PNs, each site contained a minimum core set of manageable components. These
consisted of an ATM LAN and ATM terminal equipment which were operated and
managed as a CPN, and an ATM cross-connect (XC) which was operated and managed
as part of a PN.

The interconnection of the ATM XCs at each site allowed PREPARE to form an ATM backbone which was under the control of PREPARE. This interconnection was established by using the pan-European ATM pilot network (ATM Memorandum of Understanding). Figure 2.7 depicts testbed 2.

The European ATM pilot network offered only very limited access to management information. It was therefore treated solely as the medium for connecting the networks at each site which, since they were under the control of project partners, contained network components that could be managed by the PREPARE management systems (TMN OSs).

Common to the two PuNOs was that they both owned and managed their individual ATM VP XC networks. In the PREPARE case, each of these networks in fact consisted of only two VP XCs which could be regarded as networks over which each PuNO could offer VP services in its respective geographical domain.

The management capabilities of these parts of testbed 2 varied (which reflected a realistic situation in itself). However, the communications management demonstrator developed for testbed 1 was used as the basis for testbed 2's communications management demonstrator. In particular, the management systems for the ATM components in the first PREPARE testbed were largely reused for the ATM XCs and ATM LANs.

2.8.3   The TMN Multi-Domain Management System Architecture

In the second phase PREPARE was able to demonstrate service layer interworking and management cooperation between multiple VASPs. This could be done either in a peer-to-peer relation or in a hierarchical relation as explained earlier.

Following the same approach as for phase 1, where the organisational situation was mapped directly onto a TMN-based management architecture, the organisational relations in the second phase led to the following management interfaces being investigated:

- *PuNO-to-PuNO TMN interactions.* At the level of public network operators (PuNOs) end-to-end service management takes place on a peer-to-peer basis between autonomous organisations. The networks are interconnected in order to enable and establish a *pan-European broadband infrastructure* (ATM VP XC network). The management interactions include the exchange of information reflecting network-specific service capabilities needed for various management services, such as efficient virtual path identifier management.

- *VASP-to-VASP TMN interactions.* The most likely form of such interactions is a VASP providing a customer with a service, such as a multimedia teleservice, that is integrated with a separate service, such as a VPN service, provided by another VASP. An alternative scenario is a multimedia collaboration service provider using the services of a video data base service provider. Whether the combined services will be provided to the customer through a single VASP or whether both VASPs will require relationships with the customer also needed to be investigated. A specific aspect of such types of interaction occurs in the case of:

  - *Stakeholders with multiple roles.* This covers situations where a customer with CPNs serviced by one VASP is also the supplier of a value-added service either to its own sites or to other customers' sites. It may in addition cover situations where a division of a PuNO offers value-added services but also shares some management functions with the bearer service provider division of the PuNO.

- *Customer to multiple VASP TMN interactions*. This covers the situation in which a customer subscribes to several value-added services provided by a number of individual VASPs, and accordingly the customer's S_OS needs to interwork with several VASPs' S_OSs in parallel.

Investigations into these relationships examined both peer-to-peer and hierarchical cooperation schemes as well as combinations of these schemes. Examples included new value-added services built on top of the existing ones, global VPN service provisioning and peer cooperation between providers of bearer services (ATM VP service).

Figure 2.8 illustrates the management architecture of the second phase. As in the first phase, each organisation constituted a domain containing TMN operations systems and in some cases also networks.

In this architecture the customers subscribed to the services offered by the VASPs. Therefore, the customers' management systems interoperated with the management systems of the VASPs. The VPN VASP TMN in Figure 2.8 belonged to the VPN provider, which as in phase 1 acted on behalf of the providers of bearer (ATM VP) services.

The VPN VASP S_OS interacted with only one PuNO S_OS, which in turn acted on behalf of all the PuNO TMNs involved, thus providing pan-European ATM services with single-point-of-contact features similar to those provided by phase 1's VPN VASP. Among the customers of the VPN service was one VASP which also owned network components connected to the public network. This VASP used its network to provide an information service, delivered over the VPN that was already available for other end customer organisations.

The management platforms used in phase 2 were as follows:

The Q adaptors for the ATM XCs were implemented either on the Q3ADE platform (from UH Consulting) and or on the TMN Support Facility for AIX (from IBM), depending on the manufacturer of the individual XC. The network layer OS for the PuNO TMNs was implemented on the TMN Support Facility for AIX whereas the workstations providing network operations staff with access to ATM network element management (and which therefore interfaced to the network layer OSs) were implemented on the TeMIP platform (from Digital Equipment Corporation). The service layer OSs for the PuNO TMNs were implemented on OpenView (from Hewlett-Packard).

The VPN OS in the VPN VASP's TMN was implemented on the TMOS platform (from Ericsson-Hewlett Packard Telecommunications).

The CPN service layer OSs and the multimedia conference and multimedia mail service OSs were all implemented on the OSIMIS platform (from University College London). The network layer OSs in the customer domain and in the multimedia mail service provider's domain were implemented by the use of the OSIMIS proxy agent utility [McCarthy] [NMF-026].

The IDMIS component was also used in the second phase. This time, however, it was an enhanced version which, in addition to the basic function as an integrated service interface to managed objects and directory objects, also offered a transaction functionality for managing transactions over the directory and a set of managed systems (See Appendix C).

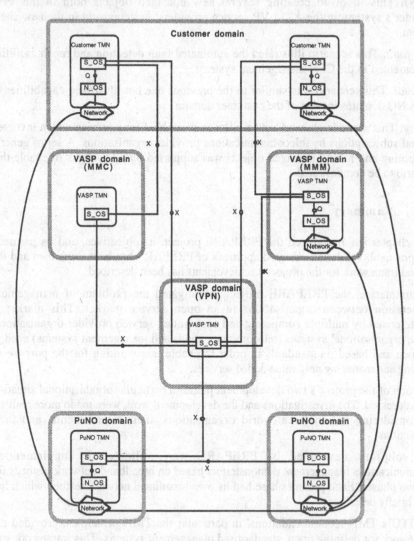

Figure 2.8: Management architecture for phase 2

The management systems shown in the architecture in Figure 2.8 were demonstrated in November 1995. At the demonstration scenarios illustrated multi-domain fault management, configuration management, and accounting management. The scenarios included the following:

Customer subscribes to teleservices for two sites. This scenario illustrated how the X.500 directory was used as the repository of service information, enabling a high degree of automation in browsing and locating service offers and in subscribing to services.

Customer adds a site. In this scenario the advanced configuration management services were utilised to reconfigure an existing VPN by adding an additional customer site (CPN). This involved creating several new managed objects both in the VPN provider's system, in the ATM VP service providers' systems, and in the new site's system.

CPN fault. This scenario illustrated the automated fault detection and repair facilities implemented in the CPN management systems.

PN fault. This scenario was similar to the previous one but illustrated capabilities in the PuNO domains instead of the customer domain.

Billing. This scenario showed how a bill was assembled for a service which involved several subscriptions by telecommunications provider organisations. A set of generic accounting and billing managed objects was supported by all systems to enable this scenario to be demonstrated.

2.9   Summary

This chapter has introduced the PREPARE project, its objectives, and its partners. The practical, validation-oriented approach of PREPARE has been explained and the general framework for the project's achievements has been described.

To summarise, the PREPARE project investigated the problem of management cooperation between organisations in an open service market. This market is characterised by multiple competing and cooperating service provider organisations. These organisations' systems (telecommunications and management systems) need to be open and based on standards in order to enable interworking for the purpose of creating and managing new value-added services.

For each of the project's two development phases a particular organisational situation was addressed. The investigations and the development work were made more realistic by considering notional real-world organisations, their requirements, and their interactions.

The solutions developed by PREPARE were validated by implementing communications management demonstrators based on broadband network testbeds for the two phases. Each project phase had its own broadband network testbed which has been briefly described.

The ITU's TMN recommendations, in particular the TMN principles, provided the framework for defining open, standardised management systems. This framework was adopted by PREPARE and applied throughout the development phases. The TMN-based management architectures developed and the management functional areas which were addressed have been briefly introduced together with the services to be managed on an end-to-end basis and the relationships between the various stakeholders involved in operating, maintaining, and managing these services.

3   From Modelling to Implementation

3.1   Introduction

The complexity of management systems such as those considered by PREPARE and the large number of key stakeholders involved in realising the services (network providers, third party service providers, third party software developers, customers, and end users) make it clear that a development methodology to support the full design and implementation cycles of the service is required. In order to provide some insight into how to address the problems of multi-domain management in future IBC networks, this chapter presents an overview of the approach taken by PREPARE when realising our communications management demonstrators.

In addition to outlining a multi-domain management system development methodology, this chapter also provides the structure to which the service examples described in chapter 4 conform.

In section 3.2 we give a brief overview of the approach taken for phase 1 of PREPARE, the conclusions we were able to extract from applying it, and how it influenced the approach taken for the second phase of PREPARE. In sections 3.3 to 3.6 we describe the methodological aspects of the system development in terms of *Requirements Definition and Analysis, Information Specification, Functional Specification and Implementation,* and *Testing and Integration,* as they were generalised from our combined experience of the two phases. Finally, section 3.7 discusses this methodological approach and draws some conclusions.

3.2   Background

The content of this chapter is based on the experience obtained from the two phases of PREPARE. Experience was gained both in understanding the problem space (i.e., TMN-based management of services over multiple domains), the solution space (i.e., the set of specifications which constituted a full system specification), and the possible interdependencies of these specifications.

The development process comprised both high-level problems of business, market, and political forces, and stakeholder requirements, as well as low-level problems of technology limitations and possibilities. To reconcile these wide-ranging influences we gradually developed an approach which successfully combined traditional top-down and bottom-up approaches until in the final stages of the project we had a good understanding of problems and solutions and how to define them.

Traditional systems development is thought of as being carried out in three main stages: requirements definition and analysis, specification and design, and implementation and testing. Although we follow this as an overall structure in the following description of our methodology there is no sharp delineation of the stages and they often overlap. The aim of this chapter is not to prescribe a complete methodology for multi-domain management system development but to describe the aspects of a methodology we believe are useful in the hope that some of them may prove helpful to the reader.

3.2.1   Development Approach in Phase 1

The approach to be adopted for the design and implementation of the PREPARE
multi-domain management system was not well-defined at the beginning of the project
since the problems of multi-domain management were relatively new ones for which
the project had to develop a common understanding. There existed no standardised
methodologies for multi-domain management system development which integrated
the service specification and the design and implementation phases of multi-domain
management systems, i.e., the complete system development life-cycle which was
required for the project.

At the outset a broad plan did exist which involved the following stages:

1. The definition of the management scenarios we wished to demonstrate, together
   with the supporting TMN architecture, management services definition, and
   information models.

2. The implementation of the intra-domain management systems required to
   manage the individual networks. This comprised the broadband network testbed
   and the implementation of inter-domain management system building blocks.

3. The testing of the inter-domain management system building blocks and their
   integration with the intra-domain management system building blocks and the
   actual broadband network testbed.

A more detailed summary of each of these stages is given in the following sections.

3.2.1.1   First Stage of Phase 1

For the first stage, four working groups were formed:

- *Scenarios group.* The aim of this group was to produce a set of management
  scenarios to reflect a representative subset of the management functionality to
  be demonstrated (such as a customer's subscription to a service, a service user's
  reservation and allocation of communications capacity, or bandwidth
  renegotiation) in a realistic organisational situation.

- *Architecture group.* This group focused on specifying an implementable TMN
  architecture for interfacing the inter-domain management system building blocks
  in each domain (see Figure 2.5).

- *Management services group.* Here the work was to define a set of management
  services to be exchanged between the operations systems (OSs) of the different
  TMNs (for example, reserve and allocate virtual private network (VPN)
  communications capacity, modify bandwidth, etc.).

- *Information modelling group.* This group worked on the definition of
  management information models required for the different OSs involved in inter-
  domain management according to the GDMO recommendation [X.722] (e.g.,
  ATM virtual path (VP) service management information model, VPN and
  DQDB MAN-based CBDS management information models).

These working groups operated largely in parallel and after running for a year the
following assessment was made of their respective output. The output from the
scenarios group described the roles of the human users and organisations involved in
the VPN service as well as the motivation for management cooperation between the

defined organisations. This was supplemented by documentation of the commercial service that the VPN provider should provide to its customers. The architecture group identified all the management building blocks required for the intended VPN services and the different management interfaces required within a TMN framework. It was generally agreed that the scenarios contributed greatly to everyone's understanding of the problem and that the architecture was a suitable basis for the implementation of the VPN service.

Management services were defined in order to specify the operations to be made available by TMN building blocks across X interfaces. Starting from a relatively informal definition some effort was made to formalise these management service specifications[1]. However, it became clear that the parallel work on specifying management information models to some extent overlapped with the management service specification work, which aimed at a similar formalised specification of operations over X interfaces based on GDMO specifications. At the same time, the management information model specifications were neither comprehensive nor detailed enough to support the defined scenarios. The management information models were general and addressed higher-level requirements rather than the requirements of the actual service examples for which the scenarios were defined.

3.2.1.2   Second Stage of Phase 1

Based on the above assessment, no further definition of the management services was undertaken and work concentrated on refining the scenarios and the management information models. The existing scenarios were therefore refined from a level where they described organisations and roles of humans to a state where the same scenarios were described in terms of operations system functions (OSFs) with detailed descriptions of the management information flowing between the OSFs. For example, reservation of VPN communications capacity involved the exchange of management information between the VPN customer's service layer operations system (OS) and the VPN provider's service layer OS, and between the VPN provider's service layer OS and the ATM VP service provider's service layer OS. Adopting this technique, a full GDMO specification of the management information models for the whole multi-domain management system was quickly arrived at. This approach also had the intrinsic advantage of ensuring that all management information modelling was directly focused on the desired implementation areas and provided a relatively brief description of the functionality associated with the management information model because each scenario described how the management information models were used to obtain a specific goal.

3.2.1.3   Third Stage of Phase 1

When the management information models became stable, the development team produced, as part of stage 3, *Test Design Specifications* (TDSs) according to the IEEE standard 829-1983 [IEEE829]. The TDSs described all the tests that involved TMN building blocks from two or more partners. They were derived directly from the information flows specified in the second stage and refined to a level of detail defining

---

1   In the first project phase management services were specified in the Abstract Service Definition Convention syntax (ASDC) defined in ITU recommendation *Message Handling Systems: Abstract Service Definition Conventions* [X.407].

the CMIP [X.711] operations between the OSs and all the parameters of the information to be exchanged. The TDSs were then used to closely guide the testing and integration of separately implemented TMN building blocks. The fact that all this detail was specified before the implementation proceeded too far was of great importance because it provided the opportunity to address and resolve many implementation-related issues that had not been apparent during the early stages of the design.

3.2.1.4   Conclusions from Phase 1

From the perspective of usefulness of the specifications produced in the first phase the following were found to best support the actual implementation:

* *Definition of Scenarios*. This clarified the aims of the whole development and the relationships between the different organisations in the organisational situation.

* *Definition of the Management Architecture*. This provided a structure for the overall multi-domain management system and for identifying the interfaces needing to be specified between the TMN building blocks.

* *Information Flows and Management Information Models*. These outlined the interactions that take place between the OSs to provide inter-domain management as well as the information visible at the interfaces of each TMN building block for inter-domain management purposes.

* *Test Design Specifications*. These described in detail the lower-level management interactions between OSs in terms of the information to be exchanged between them during the execution of the scenarios, and the sequencing of the information exchange.

It should be emphasised that these specifications deal mostly with inter-domain management interactions between the TMN building blocks. Intra-domain interactions were handled within the individual partner organisation that was implementing the TMN for the domain represented by the partner.

3.2.2   Development Approach in Phase 2

Phase 2 of the project differed from phase 1 in that the organisational situation was more complicated as it involved more service providers and because the relationships between service providers and the range of management scenarios being addressed was more ambitious. There was also a change in emphasis from just producing service, network, and network element layer OSs in each domain to adding workstation functions (WSFs) with rich functionality for human service and network administrators in each domain. Another important difference between the two phases was that while the broadband network testbed in the first phase was heterogeneous (in terms of network technologies: DQDB, ATM, Token Ring) the broadband network testbed in the second phase was homogenous (i.e., all networks were based on ATM technology) but geographically distributed.

As in the first phase the work in the second phase was roughly divided into three stages: design, implementation, and testing. Generally, the differences between the way in which these development stages were carried out in the two phases can be characterised as follows:

- *Stage 1*. The design approach was re-engineered, based on the experience of the first phase and additional specifications which were felt to be necessary to deal with the more complex organisational situation.

- *Stage 2*. The implementation work in phase 2 was distributed to a greater extent than was the case in the first phase since individual OSs rather than entire TMNs were to be implemented separately by partners.

- *Stage 3*. The testing was geographically distributed in the second phase, whereas in the first phase it was centralised and took place at the location where the broadband network testbed was installed.

3.2.2.1   First Stage of Phase 2

Based largely on the experiences of the first phase, it was felt that a more cohesive working approach was required. All the analysis, specification, design, and implementation was therefore performed in one homogenous group which would split into subgroups at various stages to address clearly-defined scenarios rather than splitting into groups addressing the different aspects of the development process. As in phase 1 the approach taken again had to combine both top-down and bottom-up aspects, although in this case more attention was paid to some of the top-down design aspects due to the increased complexity of the organisational situation. The work can be broken down into the following areas:

- *Organisational Modelling and Scenario Description*. This laid out the organisational situation on which the work of this phase was based and defined, in the form of scenarios, and the scope of the management functionality to be addressed within this context (i.e., the areas of fault management, configuration management, and accounting/billing management).

- *Role Specifications*. These provided a method for describing in more detail the requirements of the organisations involved through the detailed definition of responsibilities within the organisational situation and a way of mapping these requirements to lower-level management function requirements. As an example, several roles were identified with the VPN stakeholders which illuminated various functional management areas through the definition of financial and operational responsibilities (see chapter 4).

- *Management Architecture Design*. This provided the physical architecture of the communications management demonstrator that would constitute the framework for the more detailed design and implementation work (see Figure 2.8).

- *Information Modelling and Information Flows*. This involved identifying the management information required to perform the management functions, their definition as information models representing manageable resources within each domain, and the definition of inter-domain management operations performed on the objects in these management information models in support of the management functions required by the scenarios (for example, management information models for the VPN service, the ATM VP service, and the multimedia mail and multimedia conferencing services, as described in chapter 4).

It should be made clear that these various areas were not addressed sequentially but were to an extent interleaved, with some being revisited after the initial work on others had provided a clearer insight into the requirements associated with them.

3.2.2.2   Second and Third Stages of Phase 2

There were no major differences between stages 2 and 3 in the respective project phases concerning the development of the inter-domain management TMN building blocks. However, the stage 2 specifications needed to be even more precise and explicit about managed object behaviour because the implementers were not co-located, which made a clear interpretation of the specifications more crucial. In addition, restrictions were placed on implementers and testers for the stage 3 activities because of the new types of interdependency between the TMN building blocks which were now distributed over three sites in Europe.

3.2.3   Methodology Overview

A set of methodological guidelines was extracted from the combined experience of the two development phases. These guidelines are presented below in the form of a set of interrelated specification tasks, their main characteristics and, where available, a set of guidelines and principles. Examples for each task are to be found in chapter 4.

The methodological tasks are grouped into four key topics:

- *Requirements Definition and Analysis* comprising stakeholder analysis, conceptual resource models, role specifications, functional requirements, and scenarios.

- *Information Specification* comprising the definition of an abstract (implementation-independent) management information model which is specialised into implementation-specific representations according to some guiding principles.

- *Functional Specification* comprising a TMN functional architecture and the specification of system dynamics.

- *Implementation, Integration, Testing* comprising a multitude of diverging tasks associated with the actual system implementation: TMN physical architecture design, functionality distribution principles, organisation of integration, and testing.

Figure 3.1 provides a simplified overview of the main methodological tasks and some main relationships between tasks. In practice, there is a complex web of interdependencies and there is no simple technique which prescribes a specific sequencing of these tasks. In Figure 3.1, one possible flow is indicated if one envisages a timeline starting at the top of the figure and ending at the bottom, but in reality PREPARE did not strictly follow such a timeline.

With respect to the organisation of the work processes, there was a tendency in the second project phase for the requirements definition tasks to be separated from the other tasks, which meant that not necessarily the same people had to deal with all tasks. Clearly, however, the implementation did provide a lot of feedback on the design. The following sections describe the major tasks separately and in more detail.

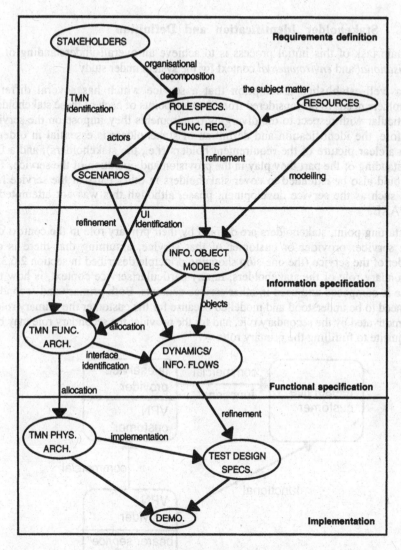

Figure 3.1: Methodology structure

3.3 Requirements Definition and Analysis

The task of defining requirements can be separated into the definition of functional and non-functional requirements. The type of non-functional requirement we are addressing here is organisational in nature in the sense that the requirements address the roles of the system under study in a multi-organisational situation. This situation provides the framework for the definition of functional requirements. Non-functional requirements, such as usability, performance, and other technology-oriented requirements, are discussed in each of the methodological tasks where they are relevant (see also section 2.6.3.2, which describes a set of non-functional service related requirements).

3.3.1   Stakeholder Identification and Definition

The main task of this initial process is to achieve an overall understanding of the *organisational* and *environmental* context for the system under study.

It is a well-established assumption that a service which has several different stakeholders needs to be considered from the viewpoints of each of these stakeholders, in particular with respect to the diverging requirements they impose on the service. Therefore, the identification and definition of stakeholders is essential in order to obtain a clear picture of the requirement holders (i.e., the stakeholders), and a first understanding of the part they play in the provision and operation of the service. This task could also be extended to cover stakeholders in other parts of the service life-cycle, such as the service development phase, although this was not attempted in PREPARE.

As a starting point, stakeholders are defined by their primary role in the context of a single service: provider or customer of that service, assuming that there is one provider of the service (the one-stop shopping principle described in section 2.6.3.2). A secondary role of the stakeholders, in any particular service context, is how that service is composed and used in other service contexts. Both primary and secondary roles need to be understood and modelled because for the customer the primary role is often motivated by the secondary role, and for the provider the secondary role may be a prerequisite to fulfilling the primary role.

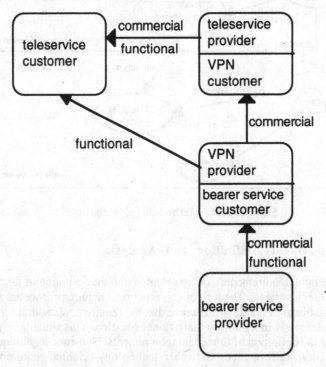

Figure 3.2: Example of management service provision in a complex value chain

For example, as depicted in Figure 3.2, the VPN service is offered by the VPN provider to the VPN customer (primary VPN stakeholders). The secondary role of the VPN customer might be that of teleservice provider, whereas the VPN provider's secondary role is that of bearer service customer. By virtue of the stakeholders' secondary roles it may also be necessary to include other stakeholders in the model in order to establish complete requirements for the service. In Figure 3.2, the bearer service provider and the teleservice customer are examples of such reference stakeholders.

Sometimes not all aspects of the service may be delivered directly by the primary provider. For example, the teleservice customer in Figure 3.2 also needs to use VPN. Therefore, it may be necessary to distinguish between two types of relationship between stakeholders: commercial relationships where a contract exists between the stakeholders, and functional relationships where one stakeholder provides services to the other stakeholder irrespective of whether a contract exists between them. The latter is the case when the primary provider subcontracts service provisioning to a third party. In Figure 3.2, the teleservice is delivered "on top of" the VPN but commercially the teleservice customer has only one relationship, namely with the teleservice provider. The VPN provider has a commercial relationship with the teleservice provider but delivers the VPN service to the teleservice customer.

3.3.2   Conceptual Models of Services and Resources

Conceptual service or resource models are conceptual models of the subject matter, i.e., the set of resources to be managed by the multi-domain management system. There is a clear need for such conceptual models as they are key both to the collaborative development effort of designers (as a universe of discourse) and to the definition of requirements of the diverse stakeholders involved. It must, however, be noted that there may not be only one such model for a particular problem area since there may be diverging views and understanding of the system depending on the roles of the persons involved: the TMN designer's view may differ from a network technology specialist's view, which again may be significantly different from a service user's view.

In PREPARE it was generally the case that conceptual resource models had to be developed in order to progress the detailed design work. This was because of the novelty of the networks and services to be managed in PREPARE's communications management demonstrators. In other cases, however, there already exist well-established conceptual resource models, some of which are internationally standardised, e.g., the ITU transmission network model [G.803]. In yet other cases such conceptual resource models are only implicitly present in a lower-level management specification, i.e., the management information model. The value of an explicit conceptual resource model should however not be underestimated.

Conceptual resource models play an important role both in the requirements definition phase and in subsequent design phases, especially in the information modelling phase where the conceptual model is somehow transformed into an object-oriented management information model.

3.3.3   Role Identification and Specification

Role specifications are a means of analysing the stakeholders' requirements on the service. They can be seen as a structured decomposition of the stakeholder model into a set of more manageable entities with a well-defined scope which are better suited for further analysis.

Role specifications comprise the identification of organisational responsibility, which is crucial in the definition of multi-domain management systems and which provides the means for analysing responsibility through decomposition into obligations, which are further decomposed into activities. Activities are expressed as functional requirements in terms of operations on the resources which constitute the system (i.e., the resources defined in the conceptual resource model described above). As an analysis tool, role specifications thus provide management functional requirements (activities) and requirements on information access modes (i.e., rights over the conceptual resources).

In a multi-domain management system it is very important to distinguish between the roles in the various domains and to establish the relationships between these roles as it is here that inter-domain management functionality is required that must be supported by the inter-domain management system. Role specifications were therefore adopted as a means of ensuring that the management functionality required by the role holders was adequately described and could be supported by the multi-domain management system being implemented.

3.3.3.1   Identification of Roles

Identification of roles in a particular context would in a real world case be based either on the actual organisational structures of the stakeholders or on a proposed future structure, depending on the required role of the system under study in the organisation.

In PREPARE we designed the organisational situations ourselves and centred role identification partly around those important aspects of the organisations which we wanted to highlight, i.e., inter-organisational relationships, and partly to achieve some level of completeness and self-containedness of the organisational situation. An example of the principles applied for the VPN and the public network operator case is described in the VPN service example in chapter 4.

3.3.3.2   Specification of Roles

A common *role specification template* was defined by PREPARE as a means of structuring in a similar manner for all roles what the actual role entails and, by refinement, in order to map from the role specification down to the operations on the conceptual resources and from management information modelling to managed objects.

The template that was developed was based on the work of the ESPRIT project ORDIT, which investigated the organisational requirements for information technology systems by examining roles and responsibilities within an organisation [StrenDob]. In PREPARE the ORDIT concepts were adopted for the specific needs of the role specification work and the particular aims of PREPARE regarding multi-domain management for the second communications management demonstrator. Therefore the role specification work was concerned with specifying the *responsibilities* of the various roles and the relationships between them according to the previously defined stakeholder model. On the basis of the responsibilities, *obligations* were described, and based on these the *activities* associated with each obligation could be described. Activities are expressed as operations on *resources*. The role specification template therefore included for each role holder the responsibilities of the role holder and to whom, the obligations that need to be discharged by the role holder in order to meet the responsibilities of the role, the activities which need to be carried out to enable the role holder to fulfil the obligations, and the *access rights* to resources required to enable the role holder to carry out the activity. (See chapter 4 for actual example role specifications.)

3.3.4   Functional Requirements

Detailed management functional requirements need to be explicitly stated in order to further the design process. Activities defined in role specifications (as defined above) provide the basis for identifying these requirements.

Various approaches to structuring functional requirements can be identified. These include the definition of management functional areas, where candidate prescribed structures are given by the Network Management Forum's *Service Management Business Process Model* [NMF-BPM] (see chapter 5), the *Telecommunications Management Functional Areas* [TMFA], the OSI Systems Management *functional areas* (fault, configuration, accounting, performance, security) [X.700], or the *TMN Management Services* [M.3200]. Observing and adopting such structuring principles in the role specification stage will significantly ease the subsequent functional requirements definition task. Requirements may also be grouped with respect to resources affected or organisational relationships (intra-organisational or inter-organisational) as defined by the role specifications.

A complementary dimension for structuring requirements is that of a life-cycle model, which provides further structure in that certain functional areas may belong mainly or only to a particular life-cycle. For example, non-functional requirements such as performance requirements may differ significantly among life-cycles. (A life-cycle approach is implicitly part of the RACE TMFAs). A further complementary approach can be to address the TMN management layer in question (see Appendix A) under the assumption that management functions are expressed differently in the different layers, in particular with respect to the level of abstraction over the real telecommunications resources involved. Also, certain management functions may be relevant to only a subset of the full range of management layers.

3.3.4.1   Scenarios

A particularly useful approach to illuminate management functional requirements was adopted by PREPARE right from the start: *scenarios* provided a means for describing the purpose of a management function in an organisational context and for defining, informally, the semantics of the function.

A scenario is defined as *an actor's interactions with the system* for the performance of a particular management function. Here, the concept of an actor is to be understood very broadly as any distinguishable part of the system and its environment which can be usefully separated out in order to study/specify the interactions of the actor with the rest of the system. Likewise, the concept of a system is to be understood as those parts of the system and its environment with which the actor interacts.

The reason for this apparently imprecise definition of concepts is that across a set of scenarios there may not be identical actors and systems since these are defined *ad hoc* for each scenario in a way which facilitates the definition of the particular goal of the scenario. In fact, individual scenarios may be (as they were in PREPARE) based on a specific view of the structure of the multi-domain management system and its environment (in terms of the communications management demonstrator and human user roles) and so specify the interactions of parts in this structure in order to obtain a given objective. This view on the system structure is typically a particular abstraction of the general TMN architecture (see chapter 2).

A scenario description consists of a set of preconditions which must be fulfilled in order for the scenario to perform. These preconditions define the view on the system, including the system structure chosen for describing the scenario, and the characteristics of each of the system's parts. Furthermore, it describes in a sequence of steps the interactions of the system's parts. Scenarios take place in a particular organisational setting (termed *organisational situation*) which is an envisaged example of a realistic market situation with customers, users, and providers of services (see chapter 2).

Scenarios are of great help for designers and implementers as an informal description technique which concentrates on *what* is taking place in the system (or between an actor, such as a management service user, and the system) rather than on *how* this is implemented. Typically, such scenarios are specified in further detail as *information flows* which define how the management function is implemented in terms of CMIS primitives exchanged between the TMN building blocks involved (this is described in section 3.5.2)[1]. In the first phase fault and configuration management was addressed through scenarios such as *multimedia conference set-up* and *QoS renegotiation*, and in the second phase fault, performance, configuration, and accounting/billing management was addressed through scenarios such as *customer adds a VPN site* and *automatic repair of faults in the customer premises network and in the public network*.

Scenarios can be regarded as orthogonal to the more structured analysis method based on stakeholders and roles. By defining a set of key scenarios that exemplify the most important functionality required of a service it is possible to check each stage of the analysis, design, and development steps to ensure that they support the scenarios adequately and therefore remain focused on the primary goals of the multi-domain management system being developed.

3.4   Information Specification

3.4.1   Identification and Specification of Information Objects

The information specification consists of the set of information objects (IOs) modelling the conceptual resources and their relationships for the purpose of multi-

---

[1]   This kind of information flow could also be specified for other components and interfaces, such as the interactions between human users of the management services and the multi-domain management system.

domain management. In the context of collaborative multi-domain management systems development the task is to model those resources which need to be exchanged and/or manipulated over domain boundaries.

IOs represent the information and operational aspects of objects required in the multi-domain management context and so constitute the totality of the relevant global management information base (MIB) (where the global MIB comprises those IOs which are needed for inter-domain management operations and so are globally visible and accessible).

The resources being modelled include both service resources (as defined in section 3.3.2) and other types of resource as described below (in section 3.4.2).

A usual step here would be to map the identified conceptual resources onto IO classes on a one-to-one basis. Then, by identifying the possible common aspects among the defined objects it is usually possible to generalise some objects and define a set of more generic objects. These must then be specialised in accordance with the previously identified requirements.

Role specifications provide functional requirements and the essential access rights, which again influence the specification of the IO classes' constituent parts, such as attributes with access modes (e.g., GET, REPLACE) and possibly notifications (see Appendix B).

Specifically, with respect to X interface specifications (see Appendix A) the approach in PREPARE took into account the fact that service management information should represent a particular network technology at an abstract level, thus hiding to a large extent the technology and implementation-specific details not needed by management service users. Therefore, service management information was intended to be user-friendly and as technology independent and abstract as possible.

3.4.2   Types of Information Object

It was evident from the experience of PREPARE that multi-domain management requires globally or widely available management information in order to enable the stakeholders involved in inter-domain management (e.g., service customer and provider) to find out the details about each other which are needed to establish inter-domain management interactions. Managed objects and the OSI Systems Management paradigm are not well suited for that purpose because it is assumed that an a priori knowledge (*shared management knowledge*) exists between the stakeholders wishing to engage in a service interaction [X.701]. PREPARE proposed the use of the X.500 directory to solve the apparent needs [X.500]. This immediately raises the problem of how to decide which information should be implemented as directory objects and which information should be implemented as managed objects.

At the initial level of information modelling such considerations about the actual implementation and/or most suitable repository would tend to obscure the modelling process itself as initially the most important aspect is *what* to model rather than *how* to implement the modelling representation[1]. Thus, IOs are abstractions above the differences between the directory objects (DOs) and managed objects (MOs).

---

1   An associated problem is that the actual notational prescriptions already differ in accordance with the actual implementation/repository at the initial design stage, i.e., there is no standardised

Analysing the semantic differences of information entities and resources to be represented in the context of multi-domain management, distinctions can be made between the following:

- Information objects representing the management view of resources (network elements, networks, services, etc.) and referred to as resource IOs.

- Information objects representing organisations and roles in the organisational situation and referred to as organisational IOs.

- Information objects required for the operation of distributed TMNs (for instance, management knowledge, logs, performance records, etc.) and referred to as operational IOs.

Using an object-oriented approach these information categories are modelled as specialisations of the generic type information object. IOs represent entities relevant to multi-domain management and are design-level constructs relieving designers from considering implementation-specific details (such as information repositories, protocols) while observing general rules easing the transition from design through specification to implementation. IOs can also be distinguished by their dynamic behaviour, their scope of visibility and the following accessibility characteristics:

- Long-lasting data (for example, information about services, networks, network elements, communication addresses) which needs to be published globally or made available to a large sector of the operator/user community and which can therefore be administered by the X.500 directory.

- Dynamic management data which can change very rapidly (e.g., packet counters in a transport entity) and which is therefore maintained by the object, or process, in which the information is created. This data is accessed and modified via OSI Systems Management services and protocols.

- Information which needs to be visible globally or at least in a wider area (e.g., information about organisations, contacts, available services, etc.).

- Information which needs to be visible within a local area only (e.g., internal addresses, network element specifics), i.e., within a single organisation.

Taking these characteristics into account, neither TMN nor OSI Systems Management standards alone are able to fulfil all the requirements regarding the representation of different kinds of IO. They can however be readily supplemented with the X.500 directory standards[1]. The proposed solution, the Inter-Domain Management Information Service, is described in detail in Appendix C.

3.4.3   Use of Existing Class Libraries / Information Models

Another very important aspect of the management information modelling task is that of reuse of existing class libraries. Although, as stated in chapter 2, there does not exist an extensive library of pre-existing objects for modelling telecommunications

---

information specification language which is readily applicable to the various candidate information repositories (which would also include for instance virtual file stores known from FTAM [ISO-8571]). A *GDIO* (Guidelines for the Definition of Information Objects) is needed in addition to GDMO [X.722].

[1]   The implications of this for the TMN architectures are described in Appendix A.

services, a range of specifications exist for modelling networks and network elements. Some of these were used in the design of the ATM network and network element management information models in PREPARE.

Generic managed objects, as specified in the OSI Systems Management function standards (see Table B.1 in Appendix B), were used to model general aspects such as alarms and states, whereas specific models and libraries were used to model technology-specific aspects, such as ETSI's management information model for the ATM cross-connects [NA52210] and ETSI's managed object library for the ATM network layer as outlined below [NA43316].

As an example, the managed object library for the ATM network layer can be used when defining a network layer management interface for different network types (e.g., SDH Ring and ATM) [NA43316]. In PREPARE the approach adopted when using this library was as follows:

1. Define requirements on the network OS.

2. Generate scenarios that include network management services which the network OS should support.

3. Select the subset of the generic class library which contains the managed object classes relevant for items 1 and 2.

4. Modify this subset to an ATM-specific network management information model. This modification was carried out by specifying subclasses, selecting from optional information, and selecting conditional packages as necessary.

However, when defining an interface for a specific type of network it is suggested that *profiling* formats such as the Network Management Forum *Ensembles* [NMF-025] and *International Standardised Profiles* [1] [NMF-TR115] are used.

3.5   Functional Specification

The separation of information and functional specifications is non-trivial when an object-oriented specification approach is adopted for the information specification. Applying GDMO requires the specification not only of the information content but also of how it can be manipulated, i.e., a functional aspect. Functional specification in PREPARE was therefore centred around the system structure rather than the functional contents of the system functional blocks as defined in the TMN functional architecture (see Appendix A). This is further discussed in section 3.7.2.

---

[1]   A profile is a way of specifying explicitly, from different features of base standards, which functional units, protocol elements, parameters, managed objects, etc., implementations have to support in order to be conformant and to interwork. One of the goals of profiles is to reduce the numerous options left open in the base standards [NMF-TR115]. Numerous profiles have already been standardised or are in the process of being standardised by ISO. These are published as International Standardised Profiles (ISPs), for example ISO/IEC ISP 11183-3:1992, *Information technology - International Standardized Profile AOM1n - OSI Management - Management communications - Part 3: CMISE/ROSE for AOM11 - Basic Management Communications*, and ISO/IEC ISP 12060-3:1995, *Information technology - International Standardized Profile - OSI Management - Management functions - Part 3: AOM213 - Alarm reporting capabilities*.

3.5.1   TMN Functional Architecture

The design of a multi-TMN functional architecture comprises the identification of individual TMNs and the subsequent structuring of these TMNs in accordance with the principles defined for the TMN functional architecture in *Principles for a Telecommunications Management Framework* [M.3010]. This structuring consists of decomposing the overall management system into a set of functional building blocks and identifying the reference points between these function blocks. Finally, the reference points need to be specified in terms of the information models applying to them, i.e., a set of interrelated information objects.

3.5.1.1   Identification of TMNs

As discussed in section 2.6.4, multi-domain management in PREPARE is modelled in terms of interworking TMNs. This means that every stakeholder has its own TMN and that inter-domain management operations between the TMNs occur at the TMN X interfaces. We therefore identified individual TMNs within their respective organisational domains.

3.5.1.2   Structuring and Decomposition of TMNs

PREPARE interpreted the TMN Logical Layered Architecture (LLA) principle quite pragmatically by defining inter-domain management as taking place at the service layer of management. The rationale for this is that it is services that are exchanged between the stakeholders and thus the management information exchanged between their TMNs is naturally service management information. Because the TMN LLA principle states that service management takes place in the service management layer, it was straightforward to envisage a TMN functional architecture in which all inter-TMN interactions take place over x reference points between service layer operations system functions (OSFs).

The architectural approach of PREPARE included a TMN service layer OSF (S_OSF) in each of the TMNs involved. At the same time, effective management of individual networks, possibly based on various technologies such as DQDB MAN and ATM WAN, would naturally be performed by dedicated technology-specific management functions which could utilise the network-specific detailed management representations for purposes such as traffic control and performance optimisation. Such management tasks are logically located at the network layer of management in accordance with the TMN LLA principle. Therefore, in every domain containing a network (CPN or PN) the TMN of that domain additionally contained a network layer OSF (N_OSF). In the detailed design of the PREPARE testbeds such N_OSFs were assumed to be implemented in separate network operations systems (N_OSs)[1]. In addition to handling network-specific management tasks, N_OSFs also provided the S_OSFs with access to the networks.

---

[1]   This was a design choice of PREPARE. TMN recommendations do not prescribe such a system to be present since an OSF can be co-located with other OSFs in one OS implementation.

3.5.2   System Dynamics

The dynamics of the multi-domain management system was defined by information flows. An information flow describes, in a diagrammatic form, a management function implementation in terms of specific protocol message flows between TMN building blocks. Thus it describes, for example, which MOs are being managed, which operations are sent to them, and which attributes and parameters are transferred between two TMN building blocks. Management information models and flows need to be designed in an iterative fashion since information flows will identify missing information which needs to be included in the information model specification and verified through updated information flows.

Typically, execution of an inter-domain management function in the multi-domain management system involves more than two TMN building blocks. To comprehensively describe all the important interactions one can use either *extended* information flows with multiple interacting TMN building blocks or *nested* information flow diagrams, in which pointers are made to other information flows for detailing aspects of one of the TMN building block's behaviour (in terms of interactions with additional TMN building blocks). Examples of the former are given in chapter 4.

3.5.3   Identification and Design of Workstation Functions

Workstation Functions (WSFs) provide the representation of systems and subsystems considered relevant and needed by a role holder. The WSF's design depends to a large extent on TMN platform technologies as different platforms offer different levels of support for developing graphical user interfaces (GUIs) and there is no common level of GUI functionality or look and feel between them. The role specifications however provided important indications of what was to be represented on the screen (the resources the role holder manages or is aware of) and the operations available to the role holder that can be invoked on these resources (which would provide clear indications for the contents of menus associated with each resource).

3.6   Implementation, Integration, Testing

3.6.1   Physical Management Architecture Design

The management architecture referred to here is the physical architecture of the total multi-domain management system. The architecture is designed on the basis of the following principles (cf. section 3.5.1):

1. Each stakeholder has its own associated TMN.

2. Stakeholders which own physical network technology have within their TMN an OS which implements management functions (OSFs) specific to the particular network being managed by that TMN, i.e., an N_OS.

3. Each TMN has an S_OS implementing service management functions associated with that particular domain and taking part in the overall distributed multi-domain management function execution.

This means that N_OSs interoperate with the S_OSs in their own TMN (via Q3 interfaces) and that S_OSs interoperate with other S_OSs in other TMNs (via X interfaces). Management end users (e.g., CPN administrators, VPN administrators) access the pertinent management function by TMN workstations interfacing the pertinent operations system (via F interfaces, although proprietary interfaces were used in PREPARE).

Attention must be paid in this area to explicitly addressing the non-functional requirements imposed by the scope of partners' interests and the TMN platforms available to partners, and to reducing the overall complexity of the management information modelling and information flow definition tasks by minimising the number of inter-domain management interfaces involved. Significantly, by addressing these issues at the architectural stage of the design process it was found that it is subsequently easier to split the work between relatively independent groups that can work on the functionality in different areas of the physical architecture. This results in an additional principle being identified:

4. For ease of implementation as few interfaces as possible should be defined. In particular, all x reference points between OSFs belonging to the same pair of interacting TMNs are implemented as one X interface.

The choice of TMN implementation platforms is a very important issue. In PREPARE this choice was not made on an overall project basis but was left to the individual partners' decision. The result of this should not be underestimated: the absence of a common platform meant that no common high-level application programming interface (API) was available across all platform technologies. This meant that software could not be ported across the platforms. Therefore, there could be no sharing of implementation work between partners (for any particular TMN building block).

The character of the commercial TMN platforms also influenced the physical architecture adopted, which in turn had an impact on the underlying functional design. Current TMN platforms are aimed at supporting very heavyweight applications that require a lot of platform functionality, resulting in very high costs for a run-time platform. It therefore seems likely that most providers and customers will not have a large number of physically separate OSs over which to distribute the management functionality for a particular service. This also reflected the situation in the project. There was therefore little incentive to divide the management functionality of a stakeholder's part in a particular service into separate OSFs since there was little potential for gaining any operational advantage (for example, due to load balancing or robustness) from running them on different platform instances.

This reinforced the tendency to define OSFs at the level of granularity that simply reflected the division into domain-specific TMNs and the split between logical layers within the TMN rather than a more finely grained decomposition of TMN building blocks.

3.6.1.1   Principles for Allocating and Distributing Functionality

When designing the TMN functional architecture some principles of distribution transparency can be adopted from the ODP Reference Model (see chapter 5) for the specific purpose of modelling and design.

The following principles were applied in the design of TMN physical architectures:

- Management representations of resources should be kept near the actual resources.

- Functionality associated with these resources should likewise be kept near the resources.

For both, *near* is interpreted as *within the same domain*. This is a natural assumption since the resources in question are within some organisation's domain and under the administrative responsibility of this domain. Organisation-specific (management) policies would be applied within the corresponding domain (and in some cases even be the basis for defining the domain).

However, in some cases other concerns have been more important than these, and thus the actual allocation of management functionality and resource (management) representation may be different. For instance, in the PREPARE architecture the VPN management functionality associated with the PNs is located in the VPN provider's domain.

3.6.2   Integration and Testing

Remote integration of distributed software is a complex and difficult task, requiring detailed conformance specifications for the interfaces where integration occurs (TDSs in the case of PREPARE, as described in section 3.2.1.3). In integrating and testing OSs developed by different partners the following techniques were found useful:

- Separation of the testing of management information model syntax from management information model semantics, i.e., ensuring all MO syntaxes are tested before the MO operation is tested. This can be quickly performed by constructing a single MO containing all the syntaxes from the MOs that constitute the TMN interface under test and testing this first.

- Using mechanisms that can return a TMN building block in an agent role to a predefined state, from which tests can recommence after a software failure.

- Making use of dummy OSs with the correct TMN interfaces to test the operation with secondary OSs that are affected by the operations of the OSs under test. This enables single interfaces to be tested in isolation without changing the operation of the OS.

3.7   Conclusions

During the four years in which the PREPARE project was active we increased our experience of the techniques required to perform multi-domain management system development successfully and obtained a greater understanding of the specific problems invoked by the multi-domain nature of this area. Techniques of iteratively refined management information models and scenario-based information flows have proven adequate for designing simpler systems, but more complex organisational situations and wider ranges of management functionality need more attention to top-down techniques. In particular, where multiple stakeholders are involved in a service, the use of role specifications provides a suitable method for mapping the overall management functionality to individual role activities and operations system functions.

In the remaining part of this chapter we discuss our conclusions, which give some indications for the future development of this methodological approach. The final section of this chapter describes related work on defining methodologies for TMN development.

3.7.1   Role Specifications

As explained in section 3.3.3, role specifications are seen as an important and very useful task in the area of requirements definition. As part of a general top-down analytical approach, role specifications were an easy to understand and easy to use structuring principle and were developed to support exactly the tasks at hand in PREPARE. The use of role specifications in PREPARE however changed over time. Initially, it was felt that they would provide a good way of describing, in a non-technical language, the meaning of *service management* for specific services, as viewed from PREPARE's perspective. Later, they became the tool for defining requirements on various multi-domain management system building blocks, for example, functional requirements were to be fulfilled by operations systems, and the resources being manipulated as an effect of role holders performing activities were to be reflected or visualised on graphical user interfaces of TMN workstations. At that point in time, the role specification template developed in the project was only descriptive in the sense of defining a set of concepts which were found useful and adequate for role specification. It was to a large extent left to the individual designer to interpret and apply the template in a way which the designer felt was useful for the task in hand. Later, there emerged a need to discuss and reuse role specifications. That implied that a reinterpretation of role specifications had to be performed, leading to the development of a more prescriptive style in order to gain a common approach and an increased guarantee for appropriateness, completeness, and correctness across the specifications.

In all these phases it was apparent that role specification was a very useful tool which provided the means for structured description and discussion of management organisational and functional requirements. It greatly assisted in focusing debate among project members at a commonly understood and consistent level of abstraction.

As we became used to role specifications a new need however emerged: a higher level of abstraction above role specifications is needed for defining and analysing the stakeholders (see section 3.3.1). Results from the ORDIT project, which were adopted and adapted for role specifications, would be expected to provide useful insights also into this problem area as well as notational and linguistic guidelines.

3.7.2   Functional Specification

The area of functional architecture and specification has not been addressed fully in the project (see section 3.5). Issues concerning the decomposition of management layers as defined by the TMN LLA principle for multi-domain management must be dealt with in more detail if TMN building blocks of different services are to be integrated to manage value-added services and if the reuse of TMN building blocks is to be encouraged. This requires a more rigorous specification of a TMN functional architecture which would provide more guidance for the design of management software applications residing in the physical TMN building blocks. In PREPARE management functional design was embedded in the information objects and information flows.

3.7.3   TMN and Methodology

The complexity of the system and of the development environment with multiple organisations, cultures, and languages, together with the variety of expertise and interests of individual participants imply that one or more structuring principles are required. The separation into requirements definition, specification, and implementation and testing is adequate for that purpose but a further structuring of each stage is required.

The methodology presented describes what actually had to be specified in the context of the project in order to enable implementation to take place. In summary, the methodology describes an approach to *multi-organisational development of multi-service, multi-stakeholder, multi-platform systems*. It is multi-organisational in that several organisations collaborated on the development, namely the project partners; it is multi-service development in that several telecommunications services were implemented and interrelated in various ways (see chapter 4); it is multi-stakeholder in that each service was constructed, provided, and managed by an individual stakeholder; and it is multi-platform in that each TMN was implemented on its own TMN platform.

TMN architectural principles are centrally positioned within this methodological framework as the way to:

- Structure and partition management functionality (section 3.5.1) by applying the TMN functional architecture and the LLA principle.

- Define the management information model (section 3.4) by applying the TMN information architecture and allocate information objects to functional blocks, thereby also defining the reference points between the functional blocks.

- Structure and partition the physical multi-domain management system (section 3.6.1) by allocating TMN function blocks to physical TMN building blocks, by applying the TMN physical architecture, and by defining the TMN interfaces by aggregating or combining functional reference points.

The only core design specifications which are not directly prescribed in M.3010 are role specifications, information flows, and test design specifications (TDSs). However, the two latter specifications may still be considered as part of the more comprehensive set of TMN recommendations by ITU. Recommendation M.3020 *TMN Interface Specification Methodology* introduces the concept of a management function which is defined as follows: "*A TMN management function is a cooperative interaction between application processes in managing and managed systems for the management of telecommunications resources (physical and logical). This normally corresponds to (sometimes a set of very few) CMIS operations or notifications. Typically, a TMN management function is the smallest part of such a cooperative interaction.*" [M.3020/94]. In this context it should be noted that TDSs differ from information flows only in the level of detail provided.

3.7.4   Related Work

The need for a methodology to support the identification and specification of the requirements and capabilities related to the management of telecommunications networks, equipment, and services is well understood by the standards and other related bodies. The main methodologies proposed to date by standards bodies are ITU's recommendation *TMN Interface Specification Methodology* and the Network Management Forum's (NMF) *Ensemble concept*. The rest of this chapter briefly describes these methodologies, their interrelations, and their relations to PREPARE's methodology[1].

3.7.4.1   *M.3020 TMN Interface Specification Methodology*

ITU's recommendation *TMN Interface Specification Methodology* is primarily designed to aid the specification and modelling of management functionality at any well-defined TMN interface [M.3020/92]. It provides a methodology for describing functional, information, and protocol specifications for TMN interfaces. Focusing on TMN functional specifications, it specifies how to derive TMN management services, TMN management service components, and TMN management functions, from which are derived management messages and associated managed objects. A main issue is the reusability of functional and information specifications. In a revised version ITU has adopted many of the concepts developed by NMF for specifying and documenting Ensembles (as described below). This would lead to more comprehensive and self-contained definitions (or documentation) of specific management solutions given precise management problems [M.3020/94]. The revised version defines the following concepts:

- A *TMN management service* addresses, as a reference, the relevant information about telecommunications management which serves a specific management goal. It is always described from the TMN user's perception of the management requirement[2]. A TMN management service is described with a prose description [M.3200], a (TMN) management goal, a (TMN) management context, (TMN) management scenarios, and a (TMN) functional architecture:

  - *TMN management goals* are the telecommunications user's benefit obtained by carrying out management activities using TMN management services.

  - A TMN *management context* defines the environment in which TMN management services are carried out. The definition includes the description of who manages the network, what is managed in it, and how it can be managed. The TMN management context is to be described using three orthogonal components: *TMN management roles* (which define the activities which are expected of the staff/system to perform telecommunications management), *telecommunications resources* (i.e., the

---

1   The *Reference Model of Open Distributed Processing* [RM-ODP] can also be viewed as a methodological framework which is complementary to the other methodologies in that the five viewpoints (see section 5.2) may be applied in any field of application. The main role of RM-ODP in PREPARE's methodology is in the requirements definition stage: stakeholder models and role specifications are closely related to the RM-ODP enterprise viewpoint.

2   Note that a set of standard TMN management services is defined in *TMN Management Services: Overview* [M.3200] which explicitly states: *"TMN Management Services are an integral part of TMN interface specifications."*

physical or logical entities requiring management using TMN management services), and *TMN management functions* (as defined in section 3.7.3).

- A *TMN management scenario* is a set of examples of management interactions using TMN management information (schema) definitions and TMN system management services and messages (e.g., CMIS [X.710]).

- A *TMN management information schema* specifies the information model of a managed system as seen over a particular interface by a particular managing application or system.

The ITU recommendation defines a series of tasks resulting in the required specifications.

3.7.4.2   NMF Ensemble Concept

NMF's *Ensemble* approach is to select, from the pool of standards available, a solution appropriate to the specific management problem and to enhance the specifications selected with other support items (management information libraries and profiles) to produce maximum effectiveness. A recommended Ensemble template is provided in *The "Ensemble" Concepts and Format* [NMF-025]. The template prescribes, among others, the following parts of a management solution specification:

- A *management context* which describes why the Ensemble is required; the *resources* to be managed; the *management functions* to be performed; the scope of the problem to be solved; and the *management view or abstraction level* (including *roles*) from which the problem is approached (cf. the M.3020 TMN management context in section 3.7.4.1).

- A *management information model* which defines the managed objects representing the resources to be managed and their interrelations, and specifies *scenarios* which show how the identified management functions are accomplished by CMIS message exchanges with the defined managed objects (cf. the M.3020 TMN management information schema in section 3.7.4.1).

NMF provides an *Ensemble Definition Process* in which several points are identified as part of the process of defining an Ensemble. In this process NMF identifies, among others, the tasks of describing the resources to be managed, identifying the functions to be applied, and identifying the managed object definitions to be applied.

3.7.4.3   Comparison

Clearly, the standardised methodologies aim at defining comprehensive and precise interface specifications and, as stated above, the revised ITU methodology is closely related to the NMF methodology. Comparing these methodologies with PREPARE's methodology it can first be observed that PREPARE broadly addressed all aspects of systems development whereas interface specifications are only a subset. Moreover, PREPARE's methodology pays relatively more attention to formalising the requirements definition stage, both by providing more rigorously defined terminology and associated notation technique, and by offering a method for the requirements definition.

The NMF has defined an Ensemble as "*the application of an Information Model in a particular Management Context.*" [NMF-025]. This statement emphasises the fact that an Ensemble is a technical solution to a defined management problem rather than a generic management specification. By adopting the same basic concepts, future ITU interface specifications (and in particular management information models) might be expected to also be less generic and more "problem-oriented".

In PREPARE, as pointed out in section 3.7.3, the specifications needed to implement TMN interfaces were the information model specification; the information flows and TDSs; and the TMN functional and physical architectures.

Comparing the concepts from NMF and ITU with these PREPARE specification items, it is clear that Ensemble definitions and interface specifications in accordance with the revised version of M.3020 [M.3020/94] contain all the relevant information. As a documentation style, the Ensemble and M.3020 approaches are therefore appropriate and sufficient to provide the specifications needed. However, from a method point of view it is less obvious whether the two standardised methodologies are appropriate. For this purpose PREPARE defined and/or applied complementary techniques, most notably role specifications and general information object specification. These lend themselves very much to the RM-ODP viewpoint approach (see section 5.2). This is consistent with the view of the RM-ODP as a *meta-architecture* which can be applied in a broad range of systems development approaches but which needs to be specialised to adapt to the specific constraints in a given field of application. In addition, the specification of very detailed TDSs is not prescribed in either of the standard methodologies. On the other hand, the definition of a TMN management function (see section 3.7.3) is broad enough to allow the view that a TDS is in fact one way of specifying TMN management functions.

4 Examples of Service Management

4.1 Introduction

This chapter consolidates the work of the previous chapters by clarifying some of the issues in multi-domain management. In this chapter are described some general inter-domain management services, some specific inter-domain service management examples, and an application example, and it is shown how these services can be combined in a specific scenario. The example service management descriptions are the result of the application of the methodologies described in chapter 3. The application example is the broadband network testbed assembled for phase 2 of PREPARE as described in chapter 2.

The services managed by these examples are an ATM Virtual Path (VP) service, a Virtual Private Network (VPN) service, a Multimedia Mail (MMM) service, and a Multimedia Conferencing (MMC) service. For each service the enterprise model providing the requirements for the services managing that service is given in terms of the stakeholders involved; the high-level resource model that the requirements are based on; examples of more detailed role specifications; and examples of functional management requirements. The design of the service management services is then given as an information model, a computational model, and a dynamic model (in terms of inter-domain information flows). The same structure is also used in presenting some general management issues that were found to be common to the inter-domain management services examined and which were therefore addressed by a common set of solutions.

Having described these services individually, the chapter then examines interdependencies and interactions of these services as an example of the complexities of designing management systems that are involved in providing and using multiple management services.

This chapter therefore provides the reader with:

- Solid examples of service management services based on actual implementations.

- An indication of the level of detail found to be required in developing management systems for inter-domain service management.

- An overview of some of the common issues relevant to inter-domain management and the implementations that address them.

- Examples of how different management services can interact to provide further value-added management services.

4.2 General Inter-Domain Service Management

4.2.1 Introduction

In this section we look at general inter-domain management services that may be applied to any specific service, including those described in the subsequent sections of this chapter. These services will be defined as a single service in the manner described in chapter 3. However, this service is intended to operate as a generic service, providing a framework of common functionality within which other management services operate. The areas addressed by this service are as follows: how to deal with multiple service subscriptions; how to deal with management services involving third parties; and how to bill for services.

4.2.2 Enterprise Model

4.2.2.1 Overall Service Description

In an open service market both consumers and providers of services will have to deal simultaneously with multiple service subscription relationships. An organisation may also be acting as both the consumer stakeholder and provider stakeholder in several service subscription relationships concurrently, for example, a service integrator. Service management services comprise part of a service subscription relationship between any two organisations.

The common view of the relationship between a service provider and a service consumer is that of the provider providing services to the consumer. When examining management aspects of services, however, it may be the case that both the consumer and the provider offer some management functionality to each other. This can be seen in a typical service subscription process where both the consumer and provider have responsibilities to each other, such as providing contact information, financial information (e.g., bank account details), etc., and later in the service life-cycle for functions such as accounting management, where the provider provides functionality in the form of supplying the bill and the consumer provides services in the form of confirming receipt and payment of the bill.

In an open service market, relationships between organisations will often not be as straightforward as the consumer-provider service subscription relationship, especially when considering the management aspects. Providers of services may often find themselves providing services to one organisation under an agreement, i.e., a contractual relationship, with another organisation. Examples of this could be:

- A public network operator (PuNO) that offers broadband wide area communications services (and corresponding management services) to an organisation which uses them as a part of some service offered by another intermediate, value-added service provider (VASP). The PuNO charges the VASP for the services and the VASP passes this cost onto its customers as part of the value-added service cost.

- A resource cost centre of a large corporation providing services to a VASP which integrates them as part of a value-added service provided to another part of the corporation. The cost centre may then bill the part of the corporation that uses the value-added service directly.

The management services described in this section constitute a basic framework upon which specific services can be built by providing a model for dealing with such multi-party service relationships. Billing services are also considered in a generic way in this model.

4.2.2.2 Service Stakeholders

Generic service stakeholders can be defined as:

- *Service consumer:* an organisation that purchases and uses a management service provided by another organisation.

- *Service provider:* an organisation that provides a management service to another organisation for which it receives payment.

- *Indirect service consumer:* an organisation that uses a management service provided by another organisation for which any payment is made to a third party organisation.

- *Indirect service provider:* an organisation that provides a management service to another organisation for which it is paid by a third party organisation.

Alternatively, stakeholders can be viewed by separating the actual service provision relationship from the commercial relationship. For the service provision relationship a service *supplier* provides a service to the service *user*. For the commercial relationship a service *vendor* provides a service to a service *customer*. The stakeholders can then be defined as follows:

- A service consumer is an organisation acting as both service user and customer with another single organisation acting as corresponding service supplier and vendor.

- A service provider is an organisation acting as both service supplier and vendor with another single organisation acting as user and customer.

- An indirect service consumer is an organisation acting as service user and customer but where the corresponding supplier and vendor are separate organisations (organisation A and organisation B respectively in Figure 4.1b).

- An indirect service provider is an organisation acting as service supplier and vendor but where the user and customer are separate organisations (organisation A and organisation B respectively in Figure 4.1c).

Figure 4.1 summarises the relationships between these stakeholders.

Suppliers, users, vendors, and customers are in themselves not very suitable for identifying as stakeholders since in an open service market no single organisation can act as any one of these alone. An organisation cannot supply or use a service without having to sell or buy it respectively, or vice versa. In other words, organisation A and organisation B as represented in Figure 4.1 cannot exist as organisations in isolation, they must be acting in some other capacity also. The service consumer, service provider, indirect service consumer, and indirect service provider are therefore the most atomic roles that can be taken by a service stakeholder.

It should be noted that the terminology used here to differentiate specific stakeholder aspects in this general model is not necessarily followed in the stakeholder descriptions throughout the rest of this chapter, where terminology more naturally suited to the specific service in question is adopted.

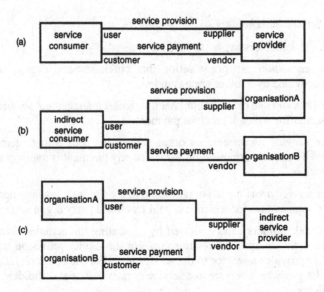

Figure 4.1: Stakeholder relationships

4.2.2.3   Service Resource Model

The resource model has to allow us to discuss the different ways in which organisations interact in the provision and consumption of services.

The basic components of such a model are the following:

- *Organisation:* an administrative unit capable of using or providing services and paying or receiving payment for such services. An organisation can itself be made up of several component organisations, although a single organisation can be contained within one parent organisation only.

- *Service instance:* a set of functions of some value that is exchanged between one organisation and another. A service instance can be broken down into several constituent service instances which may be provided by more than one organisation.

- *Service contract:* a legal agreement between two organisations giving the details of the provision of and payment for a service instance between them.

- *Bill:* a unit of payment by one organisation to another for the provision of a service.

4.2.2.4   Identification of Roles

The identification of roles is split here between those dealing with the contractual and financial aspects of service management and those dealing with the more technical and operational aspects. Stakeholders acting as service consumers have a *financial agent* role responsible broadly for locating new services, subscribing to them, paying the bills, and terminating the services. Stakeholders acting as service vendors have a financial agent role responsible for receiving requests for service subscription, granting or denying the request, sending bills, and terminating the service. Stakeholders acting as service users have a *service manager* role that deals with the operational aspects of service usage while stakeholders that act as service suppliers have a *service administrator* role that deals with the operational side of service provision. Each stakeholder also has an *owner* role to whom the stakeholder's other roles are ultimately responsible. Figure 4.2 summarises the allocation of roles to stakeholders and their relationships for the service consumer and provider situation.

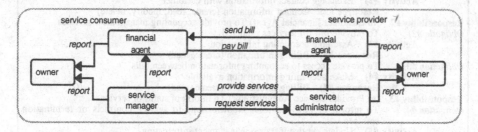

Figure 4.2: Overview of stakeholder roles

4.2.2.5   Role Specification Examples

Figure 4.3 is an example of specifications for the financial agent roles given in Figure 4.2.

The roles of the indirect service consumer and the indirect service provider will be the same as for the equivalent roles for the consumer and provider above. However the relationships to roles in other organisations will be split, i.e., for the indirect service customer the provider financial agent will belong to a vendor organisation and the provider service administrator will belong to a separate supplier organisation. Similarly for the indirect provider, the consumer financial agent will belong to a customer organisation separate from the user organisation to which the service manager role belongs.

Consumer Financial Agent

Responsibility #1) (to Owner) To deal with service related contractual and financial matters
Obligation #1)   To negotiate, establish, maintain, and terminate the contract for service instance
                with service provider
     Activity #1) Log contract information and establish repository for service instance
                      management information
          Resource #1) Contract information (create, read, modify, delete)
     Activity #2) Exchange contact information with provider
          Resource #2) Contact information (create, read, modify, delete)
Obligation #2)   To monitor and report cost of service subscription
     Activity #3) Access and log billing information
          Resource #3) Billing information (read)
Responsibility #2) (to Consumer Service Manager) To keep customer service manager informed of
                status of service
Obligation #3)   To inform of new services, modifications to service contracts, or termination
                services
     Activity #4) Update service instance management information
          Resource #4) Contract and service status information (create, read, modify,
                          delete)
Responsibility #3) (to Provider Financial Agent) To pay for service
Obligation #4)   To pay bill in timely manner
     Activity #5) Monitor bill arrival
          Resource #5) Billing information (read)
     Activity #6) Provide payment
          Resource #6) Billing information (read)

Provider Financial Agent

Responsibility #1   (to Owner) To deal with service related contractual and financial matters
Obligation #1)   To negotiate, establish, maintain, and terminate contract for service instance with
                         service consumer
        Activity #1)   Log contract information and establish repository for service instance
                         management information
                         Resource #1)   Contract information (create, read, modify, delete)
        Activity #2)   Exchange contact information with customer
                         Resource #2)   Contact information (create, read, modify, delete)
Responsibility #2   (to Customer Financial Agent) To provide accounting management services
Obligatio #2)   To provide bill for services rendered
        Activity #3)   Assemble accounting information
                         Resource #3)   Billing information (create, read)
Obligation #3)   To provide access to accounting information between bills
        Activity #4)   Make accounting information available
                         Resource #4)   Billing information (create, read)
Responsibility #3   (to Provider Service Administrator) To inform of status of service
Obligation #4)   To inform of new services, modifications to service contracts or termination
                         services
        Activity #5)   Update service instance management information
                         Resource #5)   Contract and service status information (create, read, modify,
                         delete)

Figure 4.3: Example role specifications

4.2.2.6   Management Function Requirements

This section gives an example of some of the management function requirements based on role specifications for the generic service stakeholders.

An organisation that is a stakeholder in a service must be able to locate and identify management information related to a specific service instance in a generic way. The financial agent roles for both the consumer and provider stakeholders (and their indirect counterparts) must be able to exchange and record contract-related information. It is not seen as necessary to try and model all the details of this process for use within a generic service management service, particularly since the legal ramifications of performing contract negotiation and recording the results electronically are currently not well-defined. Instead, only information that will be subsequently useful for other roles is considered here. It is therefore necessary for the financial agent to exchange and log details concerning contact information for the roles involved in consuming and providing the service as well as Service Level Agreement (SLA) information. This information needs to be accessible by the consumer service manager and the provider service administrator in order to enable direct contact between roles to deal with problems and to be able to compare the observed level of service with those agreed in the SLA for the service instance.

For accounting management the actual process of collecting usage information and assembling service charges is regarded as mostly service-specific and therefore not subject to generic service management requirements. The actual process of billing, however, is seen as an area where generic services may be applied. The functional requirements for this process are for the provider financial agent to be able to collect billing information and to send it to the consumer financial agent. The consumer financial agent has a further requirement to be able to access billing information that is correct at the time of access, regardless of when the bill was actually due.

4.2.3. Design

4.2.3.1 Information Model

As stated above, a primary requirement for the generic service management service is to be able to identify and locate management information. Here we make the basic assumption that service management information is stored and managed within the domains of individual service stakeholders. If certain information needs to be accessed by an agent (person or machine) from outside of the domain where the information is stored, then security mechanisms must be configured to enable this access and conversely to prohibit access to information not deemed to be required by that agent.

Resource	Managed Object Class
A *service instance* represents a generic (not instantiable) MO from which stakeholder-specific MOs are derived. It represents a particular instance of a service that the domain which operates the OS in which this MO exists is involved in managing. The MOs inherited from this one are placed at the head of a containment tree containing all the MOs relevant to the management of the service instance it represents.	serviceInstance (derived from "X.721:top")
A *direct consumer service instance* is a specialisation of a serviceInstance that represents a service for which the domain in which this MO exists is acting as service consumer, i.e., customer and user.	directConsumer ServiceInstance (derived from "serviceInstance")
A *direct provider service instance* is a specialisation of a serviceInstance that represents a service for which the domain in which this MO exists is acting as provider, i.e., vendor and supplier.	directProvider ServiceInstance (derived from "serviceInstance")
An *indirect consumer service instance* is a specialisation of a serviceInstance that represents a service for which the administrative domain in which this MO exists is acting as an indirect customer, i.e., as a user only.	indirectConsumer ServiceInstance (derived from "serviceInstance")
An *indirect provider service instance* is a specialisation of a serviceInstance that represents a service for which the domain in which this MO exists is acting as an indirect provider, i.e., as a supplier only.	indirectProvider ServiceInstance (derived from "serviceInstance")
A *bill* represents the payment required by the vendor of the service instance from the customer of the service instance. It is created in the vendor OS at the start date of the bill. However, the total is only calculated and payment expected from the stop date. In between these two times, billable items are included in the MO's billItemList attribute at the convenience of the vendor. The buyer may, at any time between the start and stop dates, invoke an action on the MO that forces the vendor to update the MO with all items incurred up to that point.	bill (derived from "X.721:top")
A *contact person* represents a person in another domain who plays some role in managing the service instance.	contactPerson (derived from "X.721:top")
The *consumer contact* represents a person in the domain of the consumer of the service instance represented by the MO containing this MO.	consumerContact (derived from "contactPerson")
The *provider contact* represents a person in the domain of the provider of the service instance represented by the MO containing this MO.	providerContact (derived from "contactPerson")

Table 4.1: Information model for general service management

Since in an open service market an organisation can be a stakeholder in several services and establish relationships with several other stakeholders for each instance of these services, several external agents may have requirements to access different or overlapping areas of management information. For instance, a service provider will hold management information on several different service subscriptions that only the specific consumer roles involved should be able to access.

It is clear, therefore, that the service instance should form the basis for grouping within a domain whatever information is related to that service. As the information store may be different depending on the type of service stakeholder the domain is representing, and as access to that information is based on the relationship between the stakeholder and the service stakeholder roles taken by agents in other domains, a mechanism is also needed for identifying the stakeholder type for the domain holding the management information. This in turn helps to group management information into units of security as well as aiding in the location of relevant information.

A mechanism is also required to uniquely identify a service instance in a global fashion. The mechanism suggested here is for the vendor of a service to administer a global service identifier. This is generated by concatenating a globally unique identifier for the service vendor administered by some central administration (*providerRegNum*) and an individual service instance identifier administered by the service vendor to uniquely identify all the services it sells (*contractRegNum*).

Table 4.1 gives a summary of the information model defined for general service management in PREPARE. The relationships between these MOs are summarised in Figure 4.4.

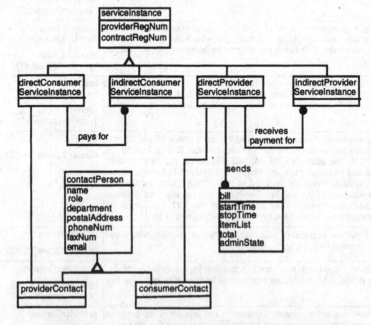

Figure 4.4: Information model relationships

This information model is not intended to cover all aspects of service management that can be generalised. It is instead intended to provide a framework information model for service management that can be used either as the basis for further generic service management information models or for the information related to specific services. It is in this latter role that this information model has been used in PREPARE. This information model primarily provides a framework for building within a particular OS containment trees that partition MOs into groups addressing the management of particular service instances. In this way, access to MOs related to a specific service instance can be conveniently restricted to the specific stakeholders and roles involved in the service.

4.2.3.2   Computational Model

In order to provide a generic definition of the management services described in this section a computational object model can be defined that is the same regardless of domain, i.e., regardless of the part played by the stakeholder and OS. Figure 4.5 gives an example interface definition for part of the computational object used to define the functionality of the contract management aspects of the general service management services. In particular, this object addresses the management and location of service instances. The example shows the interface for a consumer service manager using the ITU-T/SG15 Computational Template.

```
COMPUTATIONAL_OBJECT_CLASS       contractMgr
SERVER_INTERFACES
        NAME                     csmContractInterface
CLIENT_INTERFACES
        NAME                     statusInterface
BEHAVIOUR
END_TEMPLATE

COMPUTATIONAL_INTERFACE csmContractInterface
        OPERATION                createDirectConsumerServiceInstance
        OPERATION                deleteDirectConsumerServiceInstance
        OPERATION                createIndirectConsumerServiceInstance
        OPERATION                deleteIndirectConsumerServiceInstance
        OPERATION                locateServiceInstanceDirectProvider
        OPERATION                locateServiceInstanceIndirectProvider
BEHAVIOUR
END TEMPLATE

OPERATION createDirectConsumerServiceInstance
        INPUT PARAMETERS
                providerRegNum:      INTEGER
                contractRegNum:      INTEGER
        RAISED EXCEPTIONS
BEHAVIOUR
END_TEMPLATE
```

Figure 4.5: Example of a computational object for service instance management

4.2.3.3   Dynamic Model

Although the use of the service instance related parts of the general service management services are fairly static outside the period of service creation and termination, the billing functions do possess a more complex dynamic behaviour when operating over a multi-domain value chain of services.

Figure 4.6 shows a typical value chain situation where organisation B provides service x to organisation A by adding value to service y, which it buys from organisation C. Organisation C in turn adds value to service z that it buys from organisation D in

order to provide service y to organisation B. In each case of value being added to a
service before resale, items from the more basic service are included in the bill for the
value-added service. At point 1 in the figure organisation B reaches the end of a billing
period for service x. In order to include correct bill items relating to service y in the
service x bill, organisation B requests that organisation C updates the item in its bill
for service y. This is done by an updateBill M-ACTION request. To update its bill,
organisation C needs to include all items for service z. It therefore requests that
organisation D updates its bill for service z. Organisation D responds when it has done
this (with an updateBill M-ACTION confirm) and organisation C is able to read the
updated bill. Organisation C uses this up-to-date information on service z to update
the bill on service y, sending an updateBill M-ACTION confirm once it has done so,
thus signalling to organisation B to read the updated bill. Organisation B calculates
the bill for service x, including bill items from service y and its own value-added bill
items. It locks the administrative state of the bill, triggering an event that signals to
organisation A that this particular bill is completed and is due for payment.
Organisation A may then read the bill information to find out both the total and
itemisation details. Once the bill's administrative state is locked no further items can
be added to its item list and a new bill MO for the next billing period is created.

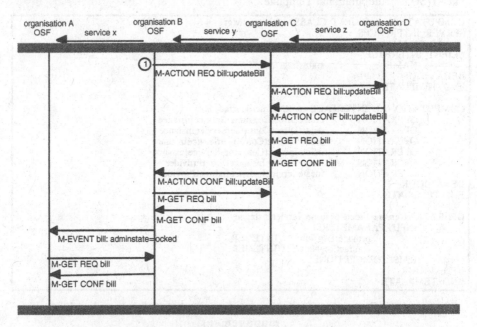

**Figure 4.6: Inter-domain management information flow for the
billing process**

4.3.   ATM Virtual Path Service Management

4.3.1   Introduction

ATM services and their management were investigated during both phases of
PREPARE. In phase 1 the ATM network was one domain in the heterogeneous

broadband network testbed. The ATM service management task was therefore to contribute the ATM part of the end-to-end services. This resulted in an ATM service management concept covering a single domain area. The emphasis in the design was on simplicity of the information model and on adopting a service-oriented approach.

In phase 2, the ATM network comprised more than one domain. The ATM service was global and could be seen as an international, multi-domain, end-to-end ATM service. This required a redesign of the ATM service management in order to be able to meet the new requirements of inter-domain management at the ATM level.

4.3.2 Enterprise Model

4.3.2.1 Overall Service Description

The telecommunications service represented by the PREPARE ATM service is a European ATM Virtual Path (VP) service, i.e., a global manageable service where it is possible to set up a VP connection between two addresses and then use this connection for the transmission of ATM cells between the access points represented by those addresses. The management service is provided through one management interface between the provider and the customer.

4.3.2.2 Service Stakeholders

In the overall enterprise model the only two stakeholders are the following:

- *ATM VP Service Customer* representing the organisation that subscribes, monitors, uses, and pays for the ATM VP service.

- *ATM VP Service Provider* representing the organisation that is paid by the ATM VP service customer to operate the ATM VP service.

These two stakeholders were sufficient for the specification and design of the management interface, but in order to ensure that the developed management concept was implementable as a system it was necessary to introduce all the related stakeholders. Therefore, the following stakeholder was also included in the enterprise model.

- *ATM VP Service Provider B* representing an organisation that is paid by ATM VP Service Provider A to provide an ATM VP service, for example, in an area not covered by Provider A.

It is here important to point out that the relationship between two providers will always be a customer-provider relationship where one provider buys a service from the other.

4.3.2.3 Service Resource Model

One of the basic characteristics of the ATM VP service is that the management of the full service is concerned with only one management interface. Therefore, the model has to include resources which represent all the necessary interchangeable information. For instance, the particular virtual path identifier (VPI) value used at an interface has to be interchanged between the customer and the provider even though it is not part of a service view.

The resources that have to be managed by this service can be grouped into four categories (see Figure 4.7):

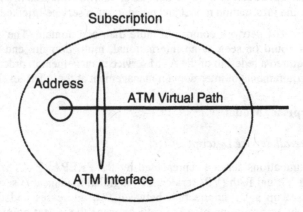

Figure 4.7: Basic ATM VP service resource model

Resources relating to ATM interfaces

Before the customer is able to benefit from the ATM VP service, it needs at least one physical interface to the public ATM network. The model for the resources related to such an interface has to give an abstract service view of the interface. However, because the interface identifies the boundary of responsibility between the customer and the provider, it has to represent the resources not only for the service view but also for the views all the way down through the physical boundary. How this group of resources is divided into more specific resources is a question of design.

Resources relating to an ATM virtual path

The basic resource for the service is an ATM VP going through the network from one end point (customer network interface) to another. This resource makes it possible for the customer to convey information in ATM cells between two points. All resources relating to a particular end-to-end VP are contained in this group. When the service is global, the VP may cross more than one domain. These groups of resources may naturally be subdivided into the basic ATM VP resource and some associated or supplementary resources (e.g., QoS or other profiles for the particular VP).

Resources relating to the addresses

Addresses are logical resources which may be used in all the domains. That means that addresses are logical service points which may contain different characteristics. The addresses are used to identify the service end points of a relation (e.g., ATM VP) as used in telephone numbers or Internet addresses. How the addresses are related to the physical resources is a design choice.

Resources related with the management service and subscription

This is a group of miscellaneous resources containing all the generally available resources which are not directly related to ATM, such as the log functionality resource and customer administration.

4.3.2.4   Identification of Roles

The identification of roles for each stakeholder in an ATM VP Service enterprise was carried out in an iterative manner during the design phase.

The ATM VP Service Customer stakeholder appears in basically two roles:

- *ATM VP Service User* utilises the ATM VP service to transmit and receive information in cells with another ATM VP service user. The user role may have different characteristics and implementations, e.g., a terminal, an application or a network.

- *Customer Service Manager* has the responsibility for the management of the service and for the general relationship to the provider.

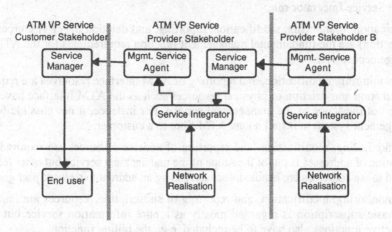

Figure 4.8: ATM VP service roles and relationships

The roles of the ATM VP Service Provider stakeholder are the following:

- *Management Service Agent* is responsible for customer care, including provision of the service, management of the service, and all other contacts with the customer.

- *ATM VP Network Realisation* is responsible for executing the transmission service in the provider's domain. It corresponds to a realisation of a network. When the provider stakeholder is the customer of another provider stakeholder, this role may also be seen as a user role.

- *Customer Service Manager*. When the provider stakeholder is acting in the Customer Service Manager role, it is because the provider is able to buy a service from another ATM VP service provider and resell the service, or parts of it, to the provider's own customers. In this case the Customer Service Manager is responsible for the service bought, its management, and all other relations with the second provider.

- *Service Integrator* has the responsibility for integrating services within the provider stakeholder. The role includes evaluation of service requests from the Management Service Agent role and subsequently giving the necessary instructions to the ATM VP Network Realisation role and the Customer

Service Manager role. The Service Integrator role is the one with the overall responsibility for the provider stakeholder.

The roles of the ATM VP Service Provider B stakeholder are the same as for all ATM VP Service Provider stakeholders (see above). Figure 4.8 gives an overview of the possible relationships between these roles.

4.3.2.5   Management Function Requirements

Management function requirements for ATM VP service management are expressed by describing the functions that have to be provided for each kind of resource. It is the particular service that is required from the Customer Service Manager role and offered by the Management Service Agent role, and this again may require the service offered by the Service Integrator role.

The creation, monitoring, modification, reporting, and deletion of VP resources (VP connection) are the fundamental management function requirements for the ATM VP management service.

The monitoring, modification, and reporting of ATM interface resources are required. The creation and deletion of physical resources such as the ATM interface have been left out of the scope of the management service. For instance, it not possible for the management system to install a new ATM fibre to a customer.

The monitoring, modification, and reporting of address resources are required. The allocation of addresses is out of the scope of the management service but other features related to the addresses are handled, such as relating an address to a closed user group.

The monitoring, modification, and reporting of subscription resources are required. The basic subscription is regarded mostly as a pure information service but more interactive functions also have to be included, e.g., the billing function.

For each resource it is also possible to separate the requirements into the five management functional areas as described in section B.4 of Appendix B, i.e., fault, configuration, accounting, performance, security. This separation did not have a direct impact on the design but it did influence the implementation priorities.

4.3.3 Design

4.3.3.1   Information Model

The information specification for the ATM VP management service was developed according to the principles described in section 3.4, taking into consideration the general requirements on an X interface, i.e., the information is structured both as directory objects (DOs) and as managed objects (MOs). The management service was therefore split into two branches: one offered by the DSF (Directory System Function, see Appendix A) and one offered by OSI Systems Management.

For the OSI Systems Management branch it was decided at an early stage to reflect the customer-provider relationship by the use of OSI manager and agent roles at the interface so that the customer is always the manager and the provider is always the agent. There are many advantages with this, for example, there were never any doubts about where MOs should be instantiated and no "handshake" problems. The task was therefore to design an information model which could be implemented in one agent.

Another early design choice was that where more than one customer was involved there would be an independent instantiation of the model for each customer. This gives maximum security for the information of any particular customer and can be based on the service instance MO described in section 4.2.

The X.500 directory service was intended to store mostly static information, both service-specific information and customer-specific information. The directory was also used for the global naming of MOs so that the whole information model could be shown in one containment tree (see Figure 4.9).

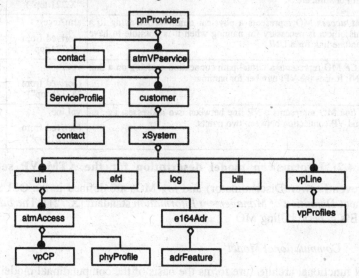

Figure 4.9: Naming tree for ATM VP management service information model

The *xSystem* object represents the root of the MO containment tree in the OSI agent, as visible to the customer manager. The *xSystem* and all the objects contained below *xSystem* are therefore MOs, whereas all the other objects are DOs.

Further specific objects may be included in the model using the same principles, with the DOs that can be used generally for all customers available at the top level, followed by the DOs for a specific customer, and then the MOs for that customer.

Resource	Object Class	Type
The *PN provider* represents the administrative information of a public network provider made available to customers.	pnProvider	DO
The *ATM VP service* DO represents the service offer information to a customer. A bearerService DO is instantiated for each service offered. In this case the atmVPservice is the bearerService.	atmVPservice	DO
A *customer* DO represents the administrative information of a customer. A customer is instantiated for each service subscription.	customer	DO
The *contact* DO represents a person to contact at an enterprise.	contact	DO
The *X system* MO represents the agent and is the root for the CMIP part of the X interface.	xSystem (derived from "X.721:system")	MO

The *uni* MO represents the service view of the UNI (User Network Interface). It contains an address MO for each E.164 address related to this interface and an atmAccess MO for each physical atm link related to the UNI.	uni (derived from "X.721:top")	MO
The *E.164 address* MO represents an E.164 number. If there are features related to the address, it contains one or more adrFeature MOs.	e164Addr (derived from "X.721:top")	MO
The *address feature* MO represents a feature which is related to the superior address MO, for example, to support Closed User Group, Free Call, Call Forward, etc.	adrFeature (derived from "X.721:top")	MO
The *ATM access* MO represents a physical atmLink belonging to a UNI. This object is necessary for naming when it is possible to have more than one link for a UNI.	atmAccess (derived from "X.721:top")	MO
The *VP CP* MO represents a virtual path connection crossing on a link at the UNI. It uses the VPI number for naming.	vpCP (derived from "X.721:top")	MO
The *VP line* MO represents a VP line between two addresses, i.e., an end-to-end VP connection between two points.	vpLine (derived from "X.721:top")	MO

Table 4.2: Information model description for the ATM VP service

The *efd* (eventForwardDiscriminator) and *log* MOs are defined in the OSI Systems Management *Definition of Management Information* standard [X.721]. The *bill* MO is PREPARE's generic billing MO.

4.3.3.2   Computational Model

The TMN functional architecture forms the basis of the computational model design. The structuring and decomposition of the TMN was carried out according to the Logical Layered Architecture (LLA) as defined in ITU recommendation M.3010 and described in section 3.5.1.2. This provides a separation into a network element layer, a network layer , and a service layer. If this is mapped to the enterprise model's roles, the service layer covers the management service agent role, the service integrator role, and the customer service manager role, while the ATM VP network realisation role is covered by the network layer and network element layer.

For the network layer and network element layer a pragmatic, straightforward approach was used for the functional model. One N_OSF is implemented per stakeholder domain with an independent workstation function (WSF) for each role. The network element layer is implemented as a Q adaptor for each piece of physical equipment, i.e., a VP cross-connect (XC).

As shown in Figure 4.10, this architecture required the definition of at least two more information models, one for the q_{atmnet} reference point and one for the q_{atmnel} reference point.

A profiling of ETSI's *TMN Generic Managed Object Library for the Network Level View* [NA43316] was used for the information model of the network layer, and ETSI's model for ATM cross-connects [NA52210] was used for the information model of the network element layer. The uses and implementation of the network and network element information models are described in [KiskSchn].

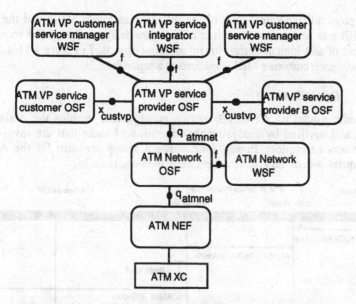

Figure 4.10: ATM VP TMN functional architecture

For the service layer a further division into functional blocks was performed. The roles that had been identified were used as the basis for this division. As the service integrator functions contain the complete management view of the service, it was natural that the service operator and the WSF interact with these functions. It was decided that the f reference point facing the WSF should also be a CMIP interface and so the service integrator role had to contain an internal information model. This model was more or less identical to the service information model at the x reference point but the internal MIB implementing this internal information had to contain information about all customers and other service information which is not relevant for customers. It therefore gives a coherent view of the service provider's entire service. This internal interface may also be used as a programming interface for management applications.

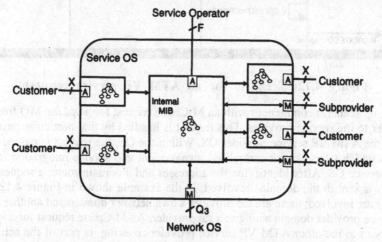

Figure 4.11: ATM VP service layer design

How the physical implementation of the different customer MIBs and the internal service MIB was realised was a matter of software design and could be more or less independent of this logical design for information objects. In Figure 4.11 a design is shown where each customer MIB is realised as a separate MIB.

4.3.3.3   Dynamic Model

The dynamic model of ATM VP service management enables the design to be exercised and verified by analysing the execution of tasks that are involved in a particular service request. Figure 4.12 shows a set-up scenario for the ATM VP service request, where the *vpLine* covers more than one domain.

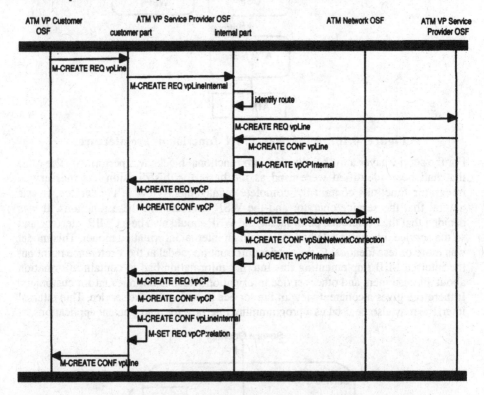

Figure 4.12: Scenario for an ATM VP service request

The set-up scenario commences with an M-Create request for a *vpLine* MO from the customer to the service provider. This request is handled by that particular customer part of the ATM VP service provider OS. Within the OS, the request is passed on to the part which takes care of centralised management and service integration for the entire service OS. After identifying the addresses and a domain route, a sequence of requests is sent to the domains involved. In the example shown in Figure 4.12, two domains are involved; these are the provider's own network domain and another ATM VP service provider domain acting as a subprovider. An M-Create request for a *vpLine* MO is sent to the other ATM VP service provider covering its part of the actual *vp* connection, and an M-Action request for a *vpSubNetworkConnection* is sent to the ATM network layer OS covering the provider's own part of the actual *vp* connection.

The requests to other domains may be carried out in parallel. When a positive response is received from these other domains, the system creates internal *vpCP* MOs representing the *vpi* numbers actually used. The *vpi* numbers have to be visible to the customer, this therefore results in the creation of a *vpCP* MO in the customer part. After this, the internal part confirms the creation of the *vpLineInternal* MO, and the customer part has to update the *vpLine* MO, which is visible to customer. It has also to set the relation from the *vpCP* to the *vpLine*; this can first be done when the *vpLine* is instantiated. Finally a confirmation is sent to the customer.

Assuming instantiated MOs before the scenario starts and focusing on the behaviour of the MOs involved, a walk through the scenario enables verification of the system design and of the information models for this particular management service.

4.4   Virtual Private Network Service Management

4.4.1 Introduction

A Virtual Private Network (VPN) service provides its customer with the ability to connect and manage remote private networks by using public IBC services efficiently so as to provide a cost saving over leased lines but with the same level of service (see chapter 2). Target scenarios for the VPN included, but were not limited to, support for the communications requirements between sites of a large multinational corporation; between internationally distributed, collaborating companies; and for value-added service providers (VASPs), such as multimedia conferencing or multimedia mail teleservice providers.

The VPN management concepts presented here are the result of iterations of analysis, design, and implementation applicable in (both technological and organisational) heterogeneous environments over the two phases of PREPARE. During phase 1 the VPN service management concept was developed to be implementable in the heterogeneous broadband network testbed comprising an ATM WAN as well as a DQDB MAN. Therefore, much attention was paid to being able to accommodate and manage this technological heterogeneity. This concept was validated in the first communications management demonstrator. In phase 2, however, the testbed was more homogenous in the sense that all network components were ATM-based. Relatively more attention was therefore given to managing the organisational heterogeneity, which was reflected in the organisational model for the second phase (see section 2.8.1).

4.4.2 Enterprise Model

4.4.2.1   Overall Service Description

As outlined in the general customer requirements, the service should provide a unified interface to the customer for several management functions, e.g., configuration, fault, performance and accounting management. For a VPN these services should be applied end-to-end, which means between terminal equipment running distributed applications communicating over the VPN. This implies that management of the customer network is also covered by the VPN management service.

The VPN model should be essentially technology-independent in that its basic design does not assume some specific underlying network technology. Where there is a requirement for the VPN service to communicate technology-specific information between domains this should be done through a well-defined specialisation of the model for the technologies concerned. This should allow the VPN service to be instantiated over different types of network technologies and, importantly, provides it with a means of instantiating a VPN over heterogeneous networks. It was also a primary aim that the abstractions developed for the service should be at a high enough level for them to offer a simple view of the management service to human users.

4.4.2.2   Service Stakeholders

The following stakeholders were identified within the overall VPN enterprise:

- *VPN Service Customer* represents the organisation that subscribes, monitors, and pays for the VPN service used by the end users.

- *VPN Service Provider* represents the organisation that is paid by the VPN customer to operate the VPN service.

- *Network Operator* represents an organisation that operates a network covering all or part of the communication paths provided by the VPN provider to the VPN customer.

Due to legislative constraints (i.e., telecommunications regulation) the Network Operator stakeholder is specialised as follows:

- *Public Network Operators (PuNOs)* are network operators that are legally obliged to provide some specific services on the open service market on conditions constrained by regulation and legislation (these are mainly the traditional monopoly telecommunications administrations to which the ONP regulations apply).

- *Private Network Operators (PrNOs)* are network operators that do not have this kind of provisioning obligation.

This distinction between types of network operator allows us to consider, in a more straightforward manner, the VPN and the CPN aspects of VPN customer organisations as two separate concerns.

4.4.2.3   Service Resource Model

To define the generic part of the VPN information model, the requirements of the non-commercial role of the various stakeholders were considered:

- The VPN customer and the VPN provider both require a single *end-to-end* abstraction for *resource reservation* and *resource allocation* between points where communications traffic will be introduced into and will exit from the VPN. This abstraction must hide the heterogeneous underlying network technology and organisational environment upon which the VPN is based. This enables the transparent migration of network technology to accommodate the changing requirements of customers, and allows for service provision on a one-stop shopping basis.

- The VPN provider requires information that shows which stakeholders are responsible for which domains.

However, the conceptual resources defined below were not developed on the basis of the above-mentioned stakeholders' requirements alone. Additional requirements (design goals) were identified:

- The model must be intuitively understandable and only reflect high-level service-oriented concerns. It should have as few concepts as possible with as few interrelations as possible, and it should abstract away from the details of specific technologies.

- The model must be applicable to a wide range of network technologies rather than provide a general concept taking all sorts of specialities into account.

- The design should also facilitate the future expansion of the service through the integration of additional service features or the replacement of existing ones by different implementations.

The concepts that were identified as being the basis of the VPN model offer two logical views on the VPN: a *connectivity* view which reflects (semi-static) topological aspects, and a *communication* view which reflects the actual communications (bit streams) running in the VPN. Two fragments of this model are defined: an end-to-end view and a domain view. Within these views virtual communication links are defined that stretch either over the entire end-to-end view or over just a single domain. To define these links we therefore need to be able to identify points on the boundaries of the end-to-end view and points on the boundaries of individual domains. These points are termed *end points*. In the end-to-end view where the end points define the boundary of the VPN as a whole, end points are named *termination points*. In the domain view where an end point marks the boundary of the domain but not the boundary of the VPN, it is named a *translation point* since it marks the point where an end-to-end communication link passes from one network's domain to an adjacent one.

The virtual communication links that constitute the VPN perform two functions. The first is the management of resource reservation; the second is the management of resource allocation. The management of resource reservation allows the customer of the VPN to request and modify the level of communication resources that is available over the VPN between two or more termination points. This therefore defines a closed group of termination points which can communicate using the VPN as well as the maximum available quality of service that can be expected between them. The unit of resource reservation between a group of termination points is called a *customer path*. The end-to-end resource reservation view can be mapped down to reserved resources in adjacent domains (operated either by PrNOs or PuNOs). A unit of domain-specific resource reservation, named a *network link,* therefore connects domain boundaries that can be either termination points or translation points.

The management of resource allocation involves specifying what resources should actually be used for communicating between two or more termination points at a designated quality of service. Resource allocation can only be performed between groups of termination points specified within a single customer path. The total allocation of resources between the group of termination points cannot exceed the level of resources reserved by a customer path between those points. The unit of end-to-end resource allocation is named a *user stream*. Depending on the underlying technology used in the individual domains the user stream is also used to convey the network interconnection requirements between the different domains that lie across its specified topology.

Additionally, customers may want to be able to optimise resource usage in order to minimise communication costs. For this purpose the concept of *scheduling* is introduced by the definition of *resource allocation profiles*, which describes how the level of allocated resources may vary over time. This enables resources to be reserved in advance of their actual allocation (for instance a user can order resources to be available for a multimedia conference the following Wednesday from 10-11 a.m.). The corresponding resource is defined as a *quality of service profile*.

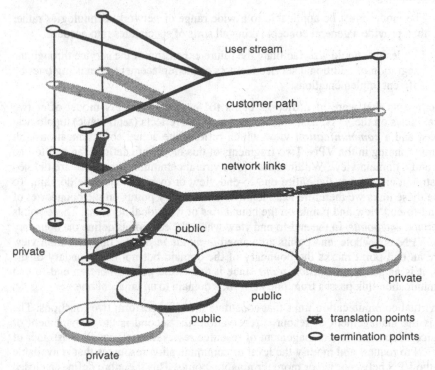

Figure 4.13: Basic VPN resource model

Customer paths, network links, and user streams can all have the following topologies:

- Unidirectional point-to-point.
- Bidirectional point-to-point.
- Unidirectional point-to-multipoint.
- Multipoint-to-multipoint, i.e., each point has an individual bidirectional link with every other point.

We group these concepts into the two *VPN views* of connectivity and communications as follows: customer paths, translation and termination points constitute the *connectivity view*, whereas user streams constitute the *communications view*. This model is sufficient for encapsulating PREPARE and technology-specific extensions to the generic VPN concepts, as discussed later. Characteristic of these two views is that resource capacity in the connectivity view is only *reserved*, whereas capacity in the communications view is actually *allocated*. Reserved means that the provider of the resource commits to provide, on request, the resource possibly at a pre-defined time. A third view could be defined as a usage view, relating to the actual use of reserved and allocated resources. However, we consider this relates closely to the signalling versus management problem which was considered out of scope for this service definition. Figure 4.13 summarises the relationships between these resource model components.

4.4.2.4   Identification of Roles

The identification of roles for each stakeholder in the VPN enterprise is based on a generic model of organisational structure where each stakeholder organisation is modelled with the following roles:

- An owner role, the holder of the ultimate legal responsibility of that stakeholder organisation, and to whom all other roles within that organisation are hierarchically subordinate.

- A financial agent which is responsible for contractual financial interactions with other stakeholder organisations.

- A manager of externally provided services (if any).

- A manager of resources managed internally within a domain, such as a network (if any).

The identification of roles in the VPN enterprise is thus as follows:

VPN Service Customer stakeholder:

- *Owner.*
- *Financial Agent* responsible for the contract with the VPN provider financial agent and for the contract with the end user financial agent.
- *VPN Service Manager* responsible for ensuring user agent requests are met by the VPN service.
- *User Agent* responsible for ensuring that users receive the quality of service they require.

VPN Service Provider stakeholder:

- *Owner.*
- *Financial Agent* responsible for the contract with the VPN customer financial agent and the contract with the network operator financial agent.
- *VPN Administrator* responsible for ensuring that the VPN customers' service requirements for provision, operation, and maintenance are met.
- *VPN Service Integrator* responsible for all management and operational interactions with individual network operator domains, and for the operation of domain technology-specific resources.

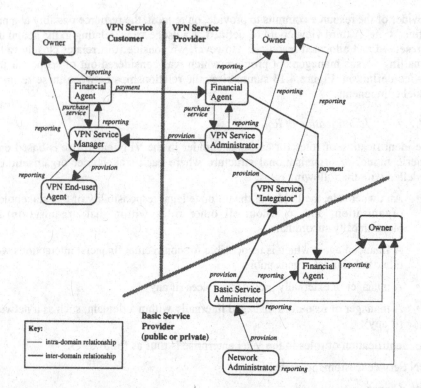

Figure 4.14: Simple VPN service enterprise role structure example

Public Network Operator stakeholder:

- *Owner.*

- *Financial Agent* responsible for billing customers (i.e., VPN provider) (and any other subnetwork providers - though this is out of scope in this context).

- *Public Basic Service Administrator* responsible for ensuring that the customer's (i.e., VPN provider's) service requirements for provision, operation, and maintenance are met.

- *Public Network Administrator* responsible for ensuring the correct functioning of the network implementing the basic service.

Private Network Operator stakeholder:

- *Owner.*

- *Financial Agent* responsible for the contract with any stakeholder using its services.

- *Private Basic Service Administrator* responsible for ensuring that the customer's (i.e. VPN provider's) service requirements for provision, operation, and maintenance are met.

- *Private Network Administrator* responsible for ensuring the correct functioning of the network implementing the basic service.

The interrelations between these roles are shown in Figure 4.14.

4.4.2.5   Role Specification Examples

The example role specifications in Figure 4.15 illustrate the definition of VPN
enterprise roles. At the lowest specification level activities are identified and expressed
partly as access modes to VPN resources. Such activities are an expression of
management functional requirements and are further specified as information flows in
section 4.4.3.3. The examples relate to VPN communication resource allocation.

VPN Service Manager

Responsibility #1) (to VPN End User Agent) To ensure that end-to-end communication paths are set
up to satisfy end user communication requirements
Obligation #1)   To ensure that end-to-end communication paths are set up between end points
associated with the VPN end users with the QoS requested by the VPN End User
Agent
    Activity #1)   Request a user stream from the VPN Service Administrator specifying the
end points and the QoS parameters
    Resource #2)   User stream (create)
    Activity #2)   Modify user streams as required by the VPN End User Agent
    Resource #1)   Termination point (read)
    Resource #2)   User stream (read, update, delete)

VPN Service Administrator

Responsibility #1) (to the VPN Service Manager) To ensure that sufficient resources have been
allocated in the VPN
Obligation #1)   To reserve requested resources in the public network operator domain
    Activity #1)   Request that a network link is reserved over the public network operator
domain.
    Resource #1)   Network Link to VPLine mapping and representation (create, read,
modify, delete)
    Resource #2)   Translation point (read)
Obligation #2)   To reserve resources in the private network domain
    Activity #2)   Request or verify that a network link is reserved over the private network
operator domain
    Resource #3)   Network Link (create, read, modify, delete)
    Resource #4)   Termination point (read)
Responsibility #3) (to the VPN Service Manager) For the end-to-end communication stream provision
and maintenance
Obligation #3)   To receive, verify, and acknowledge the request for a user steam
    Activity #3)   Verify available connectivity reservation
    Resource #3)   Network link (read)
    Activity #4)   Allocate reserved capacity in public network operator domain
    Resource #2)   Translation point (read)
    Resource #5)   User stream (create)
    Activity #5)   Allocate reserved capacity in private network operator domains
    Resource #4)   Termination point (read)
    Resource #5)   User stream (create)
Obligation #4)   To report the request for a user stream creation to customer service administrators
    Activity #6)   When subscribed to, send notifications on changes in the VPN
    Resource #6)   User stream creation notification (create)

Figure 4.15: Example role specifications

4.4.2.6   Management Function Requirements

Management function requirements are expressed by describing the functions that each
role requires from the service as a whole, derived from the activities and resources
identified for that role in the role specification. Example requirements statements
relating to the role specifications in Figure 4.15 are given here:

Create user stream

The VPN Service Manager needs to be able to request the allocation of communications resources in order to enable communications to take place within the VPN. This is done by creating user streams between termination points (which have been previously created). (Derived from VPN Service Manager Activity #1 and VPN Service Administrator Activities #4 and #5). This function is specified in detail in section 4.4.3.3.

Modify user stream

The VPN Service Manager needs to be able to modify the properties of the communications in the VPN, for instance to change bandwidth or other QoS characteristics. This is done by modifying the values of the attributes of the user stream. (Derived from VPN Service Manager Activity #2).

Report user stream creation

The VPN Service Manager needs to be informed about any communications set-up resulting in resource allocation in the VPN Service Customer's own domain. This is done by user stream creation notifications. (Derived from VPN Service Administrator Activity #6).

4.4.3 Design

4.4.3.1   Information Model

The information model part of the VPN management specification consists of the modelling of the identified resources as GDMO specified managed object classes.

Resource	Managed Object Class
A *link* represents a generic (not instantiable) communications link. The link defines the topology of the communications link which may be unidirectional point-to-point, bidirectional point-to-point, unidirectional point-to-multipoint ,or multipoint-to-multipoint.	link (derived from "X.721:top")
The *customer path* is a link that represents the end-to-end resource reservation requirements of the VPN customer. It is terminated by end points representing the edges of the customer's domain of interest (VPN).	customerPath (derived from "link")
The *network link* is a link that represents resource reservation over a domain, or set of domains, of interest. Its end points can be either the end of a chain of network links or a translation end point connecting this network link to a network link in a neighbouring domain.	networkLink (derived from "link")
The *user stream* is a link that represents the end-to-end resource allocation requirements of users in the VPN customer's domain. It defines a group of end points between which communication has been enabled at a specific QoS.	userStream (derived from "link")
The *end point* is a generic (not instantiable) representation of the points at the end of a link.	endPoint (derived from "X.721:top")
A *translation point* is an end point that additionally represents the translations from network links in the local domain to those in an adjacent one.	translationPoint (derived from "endPoint")
A *termination point* is an end point that represents the termination of a link at a point where traffic inputs to and outputs from a network application.	terminationPoint (derived from "endPoint")

The *qos spec* contains the (generic) QoS parameters to be applied to a link. Technology-specific QoS parameters are contained in MOs derived from this one.	qosSpec (derived from "X.721:top")
The *qos profile* allows the QoS applied to a link to be specified in such a way that different QoS parameters are applied at different periods of time (scheduling).	qosProfile (derived from "X.721:top")
A *connection descriptor* provides interface and routing information for a point.	connectionDescriptor (derived from "X.721:top")
A *network link relation* relates two network links. One in the local domain and one in an adjacent domain. It indicates through specialisation the maximum QoS that can be allocated over the network link via this point. A specialisation provides for different expressions of QoS suited to specific technologies and/or policies.	networkLinkRelation (derived from "X.721:top")

Table 4.3: Information model description for the VPN service

This TMN information model was developed on the basis of the VPN conceptual model presented earlier. The conceptual resources were modelled on a one-to-one basis such that each identified resource had a corresponding GDMO representation. Considering the similarities and common characteristics between some classes, the simplicity of the GDMO representation was improved by generalisation and subsequent specialisation. For instance, the previously described termination point and translation point had many aspects in common and were generalised into one end point class. Characteristics of the two different types of end point were then defined by specialisation whereby each type's particular attributes were added (see Table 4.3).

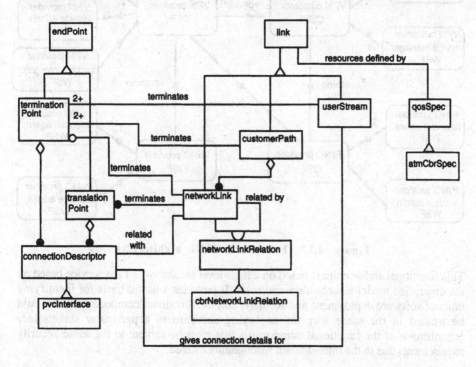

Figure 4.16: VPN information model

Technological specialisations were needed to adapt the generic objects to the ATM broadband network testbed in the second PREPARE phase. For instance, the generic *qosSpec* defining generic quality of service attributes of a VPN resource was specialised into the ATM-specific *atmCbrSpec* which added the attributes *maxBandwidth* and *cellDelayVariation*.

Figure 4.16 summarises the information model in OMT notation. It also shows (shaded boxes) where MO classes can be specialised to support specific network technologies, in this case a class that provides an ATM constant bit rate (CBR) service. The GDMO specifications of these managed object classes can be found in Appendix E.

4.4.3.2   Computational Model

To describe the computational model design it is first necessary to identify the different components of the functional architecture and the points that connect them. This is shown in Figure 4.17 as a TMN functional architecture identifying operations system functions (OSFs), the workstation functions (WSF) used by people playing roles that access the individual functions of the OSFs, and the reference points between the various functional architecture components. It is against these reference points that the different sets of management functions can be defined.

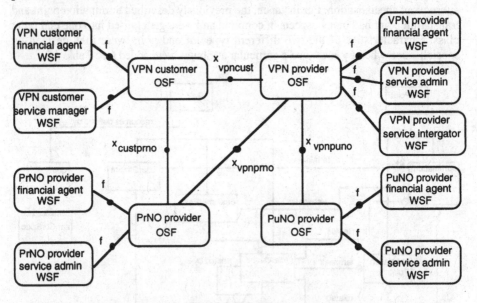

Figure 4.17: Functional TMN architecture

This functional architecture is based on a high-level breakdown of the service based on the enterprise model stakeholders and roles. It provides a useful basis for identifying units of software deployment and security, i.e., the functional components that would be treated in the same way for deploying software to a particular stakeholder organisation or the functional components that must be subject to the same security requirements due to the inter-domain functionality offered.

A further breakdown of the functional architecture into more finely grained computational objects can be performed on the basis of the functional requirements generated from the role specifications. Rather than include the entire VPN computational model we take as an example a potential unit of deployment consisting of the VPN customer OSF and the VPN customer service manager WSF. These resulted in the following areas of functionality being identified:

- End point management.
- Resource reservation.
- Resource allocation.
- Status monitoring and fault reporting.

Structuring these computational objects along these lines gives the computational model fragment shown in Figure 4.18.

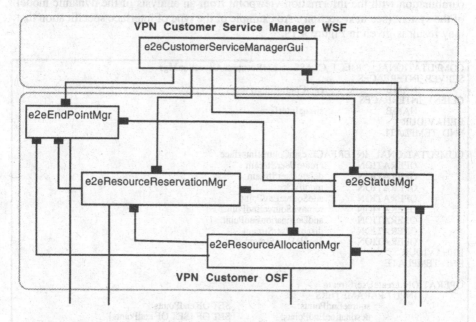

Figure 4.18: Computational model for VPN CSM WSF and VPN customer OSF

The functionality of the computational objects presented above can be summarised informally as follows:

- *e2eEndPointMgr*. Controls the creation, deletion, status, and configuration of all termination points and end points within the customer domain.

- *e2eResourceReservationMgr*. Controls the creation, deletion, and configuration of all customer paths required by the customer.

- *e2eResourceAllocationMgr*. Controls the creation, deletion, and configuration of all user streams required by the customer.

- *e2eCustomerServiceManagerGui*. This presents a graphical user interface (GUI) to a person playing the role of customer service manager (CSM). This GUI provides the CSM with an overview of the end-to-end configuration of the customer's VPN, including end points, customer paths, and user streams, and informs the user of changes to their status. The GUI also allows the CSM to manage the configuration of the end points, customer paths, and user streams that constitute the VPN by creating, deleting, or modifying them.

- *e2eStatusMgr*. This monitors the state of the customer's VPN based on information obtained from other OSFs. The current configuration of the VPN is made available to the *e2eCustomerServiceManagerGui* which it also informs of any subsequent changes. Changes to the current configuration and status of end points, customer paths, or user streams are reported to the *e2eEndPointMgr*, *e2eResourceReservationMgr*, and *e2eResourceAllocationMgr* objects respectively.

The details of the actual interfaces to these computational objects are generated in combination with the information viewpoint from an analysis of the dynamic model of the system (see next section). An example of the type of interface specification that may result is given in Figure 4.19.

```
COMPUTATIONAL_OBJECT_CLASS      e2eResourceAllocationMgr
SERVER_INTERFACES
        NAME                    csmControlInterface
CLIENT_INTERFACES
        NAME                    statusInterface
BEHAVIOUR
END_TEMPLATE

COMPUTATIONAL_INTERFACE csmControlInterface
        OPERATION               createUserStream
        OPERATION               deleteUserStream
        OPERATION               modifyQos
        OPERATION               addSourceEndPoint
        OPERATION               removeSourceEndPoint
        OPERATION               addDestinationEndPoint
        OPERATION               disableUserStream
        OPERATION               enableUserStream
BEHAVIOUR
END TEMPLATE

OPERATION createUserStream
        INPUT PARAMETERS
                sourceEndPoints:         SET OF endPoints
                destinationEndPoints:    SET OF {SET OF endPoints}
                qualityOfService:        SET OF REAL OUTPUT PARAMETERS
                userStreamId:            OBJECT IDENTIFIER
        RAISED EXCEPTIONS
BEHAVIOUR
END_TEMPLATE
```

Figure 4.19: Example computational interface

4.4.3.3   Dynamic model

The dynamic model of the VPN management service enables the information model and the computational model of the design to be exercised and verified by analysing the execution of tasks satisfying functional requirements that involve the interaction of multiple objects in a specific sequence. A comprehensive specification of the dynamic behaviour of the service systems under all possible conditions was not undertaken in PREPARE due to the scale of this task. Instead, a few selected user-level scenarios that were judged to exercise the majority of important features of the service were used to develop the dynamic model to a state where it adequately described the behaviour of the information and computation model for the purposes of implementation. Scenarios analysed were:

- Initialisation of VPN configuration by the customer.
- Reservation of resources by the customer.
- Allocation of resources by the customer.
- Addition of a further PrNO site to the customer's VPN.
- Alerting the various roles to a fault in a PrNO domain.
- Alerting the various role to a fault in a PuNO domain.

Preconditions were assigned to these scenarios and their dynamic behaviour was then examined at two levels. At one level the dynamic behaviour of OSFs was examined. This was important since these interactions occurred across the x reference point in the VPN functional architecture and, therefore, the dynamic functionality occurring over this reference point could be mapped down to information flows of CMIS primitives between these OSFs. This aided in verifying the inter-domain information model and in ensuring that the managed object classes at inter-domain interfaces satisfied the requirements imposed by the scenarios. These flows later formed the basis of the test design specifications used to test the interfaces between the implementations of the separate OSFs.

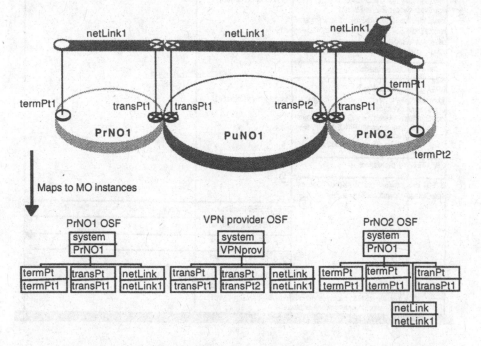

Figure 4.20: Example VPN configuration as precondition to information flow

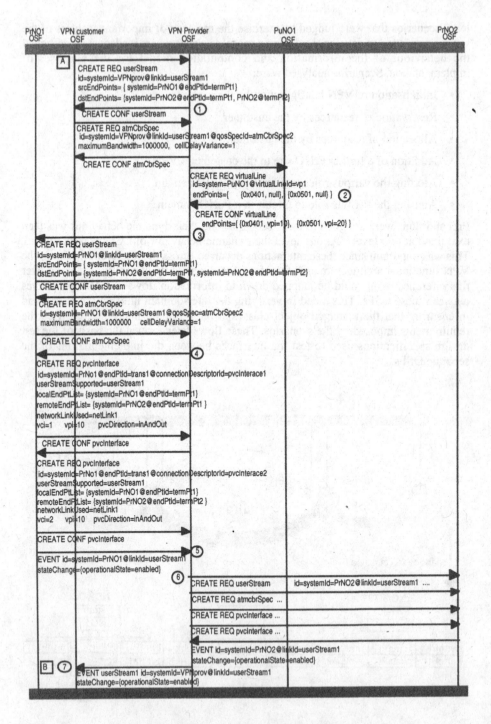

Figure 4.21: Example information flow

The other level of dynamic modelling was between computational objects within an OSF. This acted to verify the computational model and to specify the behaviour of the computational objects. These two levels of the dynamic model could be directly interfaced for any particular scenario. This aided specification of the relation between computational objects and CMIS information flows. It also further clarified the behaviour of the computational objects and ensured that engineering viewpoint considerations about how computational objects were mapped onto engineering objects interacting as part of TMN OSFs were supported.

The rest of this section provides an example of how the dynamic model for the resource allocation scenario is specified both as an inter-OSF CMIS information flow and as dynamic interactions between computational objects in a specific OSF.

The information flow described here shows how CMIS primitives are exchanged over the $X_{VPNCust}$, $X_{VPNPrNO}$, and $X_{VPNPuNO}$ interfaces. The specific situation in which this information flow occurs is summarised in the Figure 4.21. The single PuNO offers an ATM VP service (see section 4.3) over the $X_{VPNPuNO}$ interface while both PuNO1 and PuNO2 offer access to VPN services over $X_{VPNPrNO}$ interfaces. The resource allocation request is for a multipoint-to-multipoint communications link between one termination point on PrNO1 and two on PrNO2. Figure 4.20 details the specific configuration assumed at the beginning of the information flow example in Figure 4.21.

The circled numbers in Figure 4.21 mark the following actions:

1   The customer initiates resource allocation creating a *userStream* MO in the VPN provider OS with an associated *qosSpec* MO.

2   The VPN provider OS requests a *vpLine* from the PuNO OS to connect the two PrNO domains across the single public network (PN) domain that the OS is responsible for.

3   The VPN provider OS requests resource allocation in the PrNO1 domain by creating a *userStream* and a *qosSpec* MO in the PrMO1 OS.

4   The VPN provider OS provides the PrNO OS with the *pvcInterface* MOs that specify the technology-specific connection details as well as the network link to take the requested resources from.

5   Once the permanent virtual circuits (PVCs) have been set up in the PrNO domain, the PrNO OS notifies the VP provider OS that the *userStream* in its domain has an enabled operational state.

6   The VPN provider OS goes through a similar process with the PrNO2 OS.

7   Once the user streams have been set up in the PrNO OS the customer OS is notified that the user stream operational state is enabled.

The dynamic model for the part played in the resource allocation scenario by the computation objects from the VPN CSM WSF and VPN customer OSF (see previous section) is given in Figure 4.22. Note that the boxed points A and B correspond to the similarly marked points in the CMIS information flow of Figure 4.21.

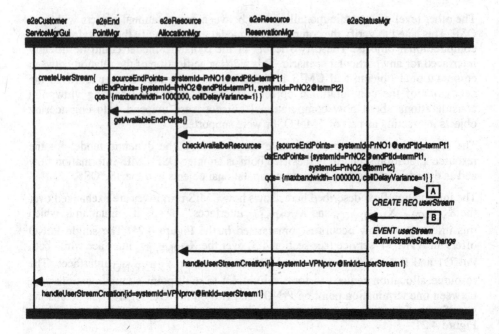

Figure 4.22: Example of information flow between computational objects

4.5 Multimedia Mail Service / Global Store Service Management

4.5.1 Introduction

The Multimedia Mail (MMM) service is designed to establish an electronic mail system for multimedia messages. It is based on the ITU recommendation for a message handling system [X.400] and defines additional body parts in the message structure. Therefore, apart from ordinary text a multimedia message can contain still and moving images as well as audio and video [Moeller].

The message handling system consists of user agents for sending and receiving messages, and intermediate message transfer agents. The message transfer agents transport a message from the originator to the recipient in a store-and-forward manner. Multimedia messages composed of audio and video claim a lot of storage capacity, which is a limited resource on a typical message transfer store. Thus an intermediate storage service is called for.

The Global Store service offers to make any data, especially high volume data, accessible worldwide. An end user can put an electronic document into the global store and receives back a reference which uniquely identifies the document. By distributing the unique reference other people can be given access to it. They can call the global store service with the unique reference and receive back the original document. The receiver may choose to receive the document directly, for example, a video clip can be played out in real-time from the global store. The receiver does not, therefore, need large amounts of free storage space.

In the multimedia mail service, high volume body parts are automatically extracted from a message, put into the global store, and replaced by the reference that is returned. Then the message is transported from the originator to the recipient where the references are automatically resolved when the message is presented.

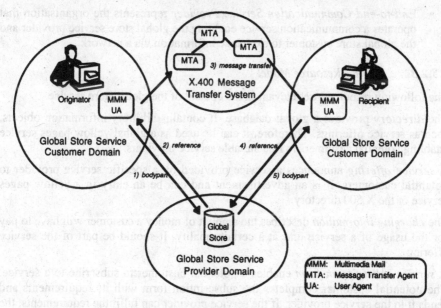

Figure 4.23: Global store service in the context of a multimedia mail service

4.5.2 Enterprise Model

4.5.2.1 Overall Service Description

To provide the global store service on a commercial basis we need additional management functionality:

- Configuration mechanisms to register and deregister customers and to set up storage media.
- Accounting mechanisms to bill customers according to their service usage.
- Fault mechanisms to supervise and maintain storage media and network components.

4.5.2.2 Service Stakeholders

The following stakeholders were identified within the global store service:

- *Global Store Service Customer* represents the organisation that subscribes, monitors, and pays for the global store service usage of the end users in the customer domain. Note that the global store service customer might also be a value-added service provider. A multimedia mail service provider, for example, can enrich its service by offering a global store to its customers as described in the usage scenario in the introduction to this section.

- *Global Store Service Provider* represents the organisation that operates the global store service and is paid by the global store customers. It is the end-to-end communication service customer and, therefore, this organisation has to pay for the communication service usage.

- *End-to-end Communication Service Provider* represents the organisation that operates a communication service enabling the global store service provider and the global store customer to exchange information via a network.

4.5.2.3   Service Resource Model

The following resources are relevant in the context of the global store service:

The *directory* provides a global database. It contains arbitrary information objects, such as service offerings. Therefore, it can be used as a global yellow pages service enabling potential customers to find suitable service providers.

A *service offering* announces a service provided by a specific service provider to potential customers. It is an advertisement and can be an entry in a yellow pages service or the X.500 directory.

The *charging information* describes the amount of money a customer will have to pay for the usage of a service unit at a certain quality. It should be part of the service offering.

A *service subscription form* enables a potential customer to subscribe to a service. The potential customer completes the subscription form with its requirements and sends it to the service provider. If the service provider can fulfil the requirements, the service provider and the customer sign a service contract.

A *service contract* fixes the rights and obligations for both the customer and the service provider.

A *customer profile* contains all customer-specific attributes associated with a service. It is created by the service provider after a service contract with the customer has been signed.

An *end user profile* contains all end user specific attributes associated with a service. For each end user within a customer domain an end user profile is created by the service provider or the customer.

The *usage information* is gathered during the operation of a service. It enables the service provider to bill its customers according to their used resources.

The service provider sends a *service bill* to the customer at regular intervals. It lists the amount of money that the customer has to pay for service usage. It should be itemised per end user.

The customer can complain about the provided quality of service by sending a *trouble ticket* to the service provider which contains a description of the problems that the customer encountered.

A *trouble report* informs the customer about problems with a service, the recovery plan and an estimated time for recovery. The service provider sends out a trouble report to the customer when it encounters problems itself or when it has received a trouble ticket from the customer.

A *document store* is a system for the long-term storage of documents, e.g., a workstation equipped with a hard disk.

The *customer premises network* includes all the terminal equipment and networking components of a customer domain.

The *public network* includes all the networking components of a public domain. Due to legislative constraints the interconnection of customer premises networks still has to be carried out via public networks in most European countries.

A *document* is a collection of information in electronic form.

4.5.2.4 Identification of Roles

The identification of roles for each stakeholder in the global store service enterprise model is based on a generic organisational structure. Figure 4.24 shows the roles that are involved in a service contract and the service usage between the global store service customer and the global store service provider.

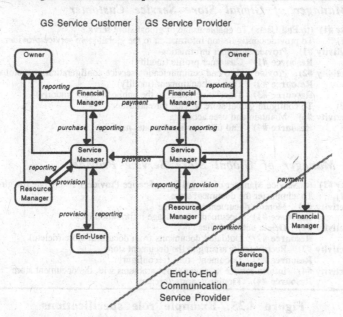

Figure 4.24: Stakeholder and role relationships

The roles identified are as follows:

- *Owner* is the holder of the ultimate legal responsibility of the stakeholder organisation to whom all other roles within the organisation are hierarchically subordinate.

- *Financial Manager* is responsible for contractual financial interactions with other stakeholder organisations.

- *Service Manager* is responsible for the contractual provision and usage of resources and services vis-à-vis other stakeholder organisations and the end users within the stakeholder organisations.

- *Resource Manager* is responsible for the provision of resources within the stakeholder organisation.

- *End User* is allowed to use a specific service.

Note that we have identified only those roles within the end-to-end communication service provider domain that interact with roles of the global store service provider domain as the end-to-end communication service customer.

4.5.2.5   Role Specification Examples

In this section the responsibilities of the roles of the global store service customer and the global store service provider domain are described. Based on these responsibilities, the roles' obligations and activities are derived and specified. One example is given for each domain and is also used to illustrate the required management functions, the user interfaces, and an information flow specification in the following sections.

Service Manager of Global Store Service Customer

Responsibility #1) (to End Users) To enable usage of global store service
Obligation #1)   To provide configuration information to the global store service provider
    Activity #1)   Provide customer information
        Resource #1)   Customer profile (modify)
    Activity #2)   Provide end-to-end communication service configuration of customer sites
        Resource #2)   Terminal equipment (modify)
        Resource #3)   Translation points to the public network (modify)
Obligation #2)   To configure global store service for end users
    Activity #3)   Maintain end user accounts
        Resource #4)   End user profiles (create, modify)

Resource Manager of Global Store Service Provider

Responsibility #1) (to Service Manager of Global Store Service Provider) To operate the global store
Obligation #1)   To administer the document store
    Activity #1)   Monitor document store usage
        Resource #1)   Document store usage information (read)
    Activity #2)   Delete outdated files
        Resource #2)   Outdated documents from document store (delete)
    Activity #3)   Rectify faults arising in the document store
        Resource #3)   Document store (reconfigure)
    Activity #4)   Inform service manager about problems with the document store
        Resource #4)   Trouble reports (create)

Figure 4.25: Example role specifications

In the PREPARE project the global store service was operated on top of the VPN service. The responsibilities resulting from a contract between the global store service provider as the end-to-end communication service customer and the VPN service provider as the end-to-end communication service provider can therefore be directly taken from section 4.4.

4.5.2.6   Management Function Requirements

Functions used by the service manager for administering end users, and functions used by the resource manager in the global store service provider domain for administering the global store were selected to provide examples of management functions supporting the roles in their activities.

Administration of End Users

The service manager needs to be able to create, modify, and delete end user accounts. This enables the service manager to give end users access to the global store service. By setting the parameters within the end users' profiles the service manager can individually set up restrictions on their service usage.

Administration of Document Store

The resource manager needs to be able to supervise the operational state of the document store and must keep track of the used storage space. When a certain threshold is exceeded, outdated documents have to be deleted. Based on the storage usage, data capacity expansion has to be planned to avoid running out of space and customers complaining about the service availability.

4.5.3 Design

4.5.3.1   Information Model

The objects in the information model for this service are described in Table 4.4 and their relationships summarised in Figure 4.26.

Resource	Object Class	Type
The *service offering* describes the characteristics of the service provider including general charging information. In addition, it gives a contact point for the subscription.	serviceOffering	DO
The *service subscription form* is not modelled as management information as other mechanisms such as electronic mail are used.	n/a	-
When the *service contract* is concluded, it is represented by a service instance object in each domain. All contract-specific information is stored below it.	serviceInstance	MO
The customer-specific *charging information* is held in an accounting profile that can set up a basic rate as well as usage-based charges.	accountingProfile	MO
The *customer profile* stores the customer-specific global store service set-up parameters.	gsCustomerProfile	MO
The *end user profile* stores the end user specific global store service set-up parameters.	gsEnduserProfile	MO
Each *document* in the global store is represented by a separate object containing the related information, e.g., the fidelity time. The object is created when a document is put into the global store and removed as soon as the document is deleted.	gsDocument	MO
To gather *usage information* the creation and deletion of gsDocument objects is logged.	logRecord	MO
The *bill* represents the payment required from the customer of the service by the provider of the service.	bill	MO
A *trouble ticket* is not modelled as management information as other mechanisms such as electronic mail are used.	n/a	-
A *trouble report* is not modelled as management information as other mechanisms such as electronic mail are used.	n/a	-
The state of the *document store* including used and free capacity, for example, is held in the gsService object.	gsService	MO

Table 4.4: Information model description for the global store service

Since the global store service is realised as a customer of the VPN service, the customer premises network resources as well as the public network resources are covered by the information model of the VPN service.

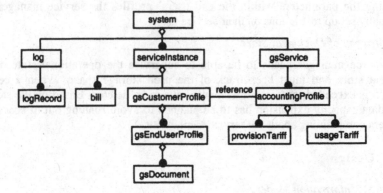

Figure 4.26: Information model relationships

4.5.3.2 *Computational Model*

The functional architecture of the system is illustrated in Figure 4.27. It consists of a central global store service OSF acting on the real resources and several workstation functions that provide access to it. The components communicate via the q and x reference points using CMIP.

The system can be broken down into small functional units. They are used to fulfil a specific group of tasks, such as end user administration, as identified when analysing the functional requirements. For the different roles in a domain it is possible to instantiate a certain set of the functional units. Thus the WSFs only obtain access to those parts of the system functionality that they need in order to perform their activities.

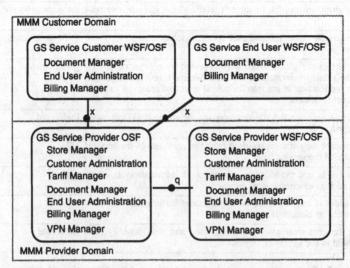

Figure 4.27: Functional architecture

4.5.3.3   User Interface

The workstation functions provide a graphical user interface based on the X Window system. To illustrate the communication of the roles with the system at the g reference point we selected as examples the document store administration (see Figure 4.28) and the user administration functionality as it is part of the customer workstation function (see Figure 4.29).

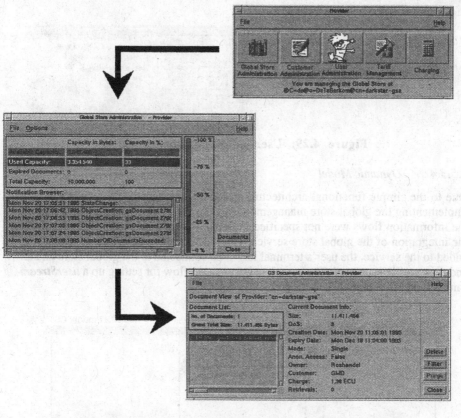

Figure 4.28: Document store administration WSF

A primary window gives access to the functional units. If the user administration is selected by the service manager a list of the registered users is presented. Using the operations create, delete, etc., the list of users can be modified. The parameters of the end user profile can be set in a separate window.

The document store administration allows the resource manager to supervise the global store, especially the used storage space. To delete documents the resource manager can move directly to the document administration.

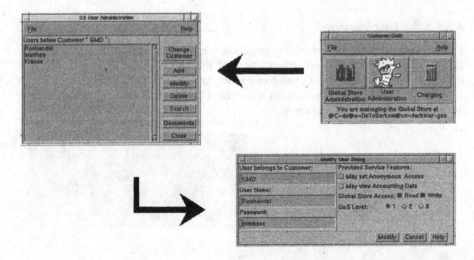

Figure 4.29: User administration WSF

4.5.3.4   Dynamic Model

Due to the simple functional architecture based on one operations system function implementing the global store management and the associated workstation functions, the information flows were not specified in detail. Figure 4.30 shows an example for the integration of the global store service with the VPN service. When an end user is added to the service, the user's terminal has to be connected to the global store and an end user profile is created. The complete information flow for setting up a *userStream* can be found in section 4.3.

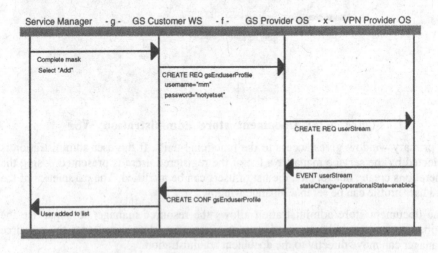

Figure 4.30: Example of an information flow

4.6 Multimedia Conferencing Service Management

4.6.1 Introduction

The Multimedia Conferencing (MMC) service is designed to support person-to-person communication between two or more people by taking advantage of the audio and video facilities and shared applications available on multimedia workstations and personal computers. To achieve multi-way communication between conference participants while incurring the minimum use of network resources, multimedia data streams are transported using Internet Protocol (IP) multicast. [Handley]

The service is intended for use by organisations wishing to enable communications between people working from remote, privately operated networks in a controlled and observable manner. Although wide area multicast services are emerging on the Internet, requirements for privacy and accountability may require service communications traffic to be transported by a private multicast network as supported by the management service described here.

4.6.2 Enterprise Model

4.6.2.1 Overall Service Description

The service consists of a private IP multicast backbone network that allows desktop multimedia conferencing applications operated by designated end users working from specific private networks to communicate. The conferencing service allows its users to request conferences between a group of named individuals over a specified set of communication media at a specified quality of service.

The service provides facilities for accounting for usage of the conference services, thus enabling usage details to be provided in an itemised bill. Faults in equipment and networks are also monitored and their effect on the service summarised for management users.

4.6.2.2 Service Stakeholders

The service stakeholders in the multimedia conferencing service enterprise model are:

- *MMC Service Customer* is the organisation that subscribes, monitors, and pays for the MMC service used by the end users.
- *MMC Service Provider* is the organisation that is paid by the MMC customer to operate the MMC service.
- *Multicast Router (MCR) Operator* is an organisation operating a multicast router.
- *Network Provider* is an organisation that provides end-to-end communication services to the MMC service provider.

4.6.2.3 Service Resource Model

The service resource model addresses two basic aspects of the service. The first is the multicast network over which the multimedia conferencing applications operate. The second deals with controlling, monitoring, and logging individual conferences.

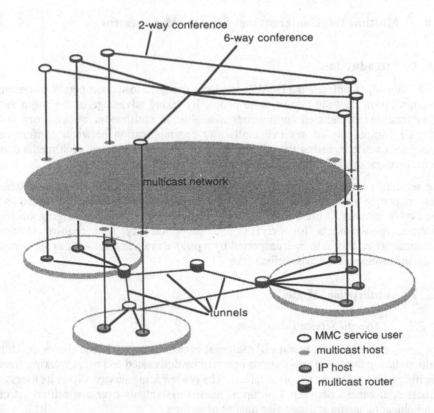

Figure 4.31: Multimedia conferencing service resource model

The model for the multicast network involves the management of *multicast routers* and their interconnection by *multicast tunnels* in order to allow multicast traffic to operate over wide area networks that support only point-to-point connections. By connecting IP hosts over a broadcast/multicast-capable LAN to a multicast router and then connecting this router by tunnels to similar ones at other user sites (via intermediary multicast routers if necessary), a multicast network is built up that provides multiway communication between all terminals attached to it. Multimedia conferences can be configured over this multicast network by service users at IP hosts. Figure 4.31 shows a possible configuration of IP hosts, multicast routers, and tunnels making up a multicast network over which different multimedia conferences are configured.

4.6.2.4   Identification of Roles

The specific roles involved in the management of the MMC service were selected along similar lines to those chosen for the other services in this chapter. As well as owner roles for each stakeholder , the following roles were identified:

MMC Service Customer

- *Financial* Agent responsible for the contract with the MMC service provider financial agent.

- *MMC Service Manager* responsible for ensuring that end user requests are met by the system.
- *MMC End User Agent* responsible for expressing end user requirements.

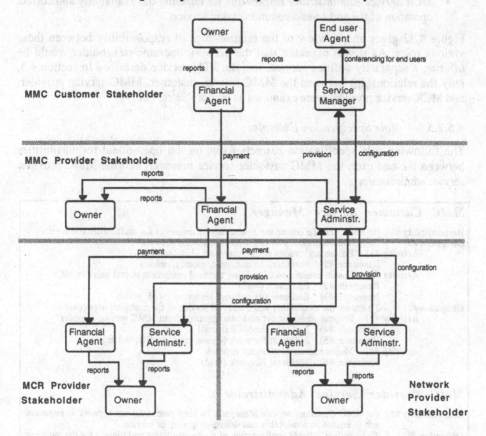

Figure 4.32: Responsibilities between stakeholder roles

MMC Service Provider

- *Financial Agent* responsible for the contract with the MMC service customer financial agent and for the contracts with the network operator and MCR service providers.
- *MMC Service Administrator* responsible for ensuring that the MMC service customer's requirements for provision, operation, and maintenance are met. This role is also responsible for integrating the operational aspects of the services used from the MCR service providers and network provider.

MCR Service Provider

- *Financial agent* responsible for the contract with the MMC service provider.
- *MCR Service Administrator* responsible for ensuring the availability and correct operation of the MCR service.

Network Provider

- *Financial agent* responsible for the contract with the MMC service provider.

- *MCR Service Administrator* responsible for ensuring the availability and correct operation of the end-to-end communication service.

Figure 4.32 gives an overview of the relationships of responsibility between these various roles. As it was expected that the network operator stakeholder would be offering a separately defined service, i.e., the VPN service described in section 4.3, only the relationships between the MMC service customer, MMC service provider, and MCR service provider were examined as part of this service.

4.6.2.5   Role Specification Examples

The following role specification extracts focus on the operational responsibilities between the end user, the MMC customer service manager, and the MMC provider service administrator.

MMC Customer Service Manager

Responsibility #1) (to End User) To ensure the availability of resources for multimedia conferences
Obligation #1)   To ensure accessibility of end users to multicast network
    Activity #1)   Ensure registration of end user with MMC service
        Resource #1)   End users (create, read, modify, delete)
    Activity #2)   Ensure connection of end user terminal equipment to end user site MCR
        Resource #2)   End users (read)
        Resource #3)   End user site network resources (read, write)
Obligation #2)   To ensure the availability and quality of service of the multicast network
    Activity #3)   Request multicast network configuration from MMC service provider
        Resource #4)   End user site MCRs (read)
        Resource #5)   Multicast Network (create, read, modify, delete)
    Activity #4)   Monitor status of multicast network
        Resource #6)   Multicast Network (read)

MMC Provider Service Administrator

Responsibility #1) (to MMC Customer Service Manager) To configure multicast network as requested and to ensure its availability and delivered quality or service
Obligation #1)   To identify a suitable configuration of multicast routers and tunnels for the multicast network that satisfies customer requirements
    Activity #1)   Match customer requirements for multicast network to the potential configuration of customer site MCRs, third party MCRs, and possible connections between them
        Resource #1)   Customer multicast network request (read)
        Resource #2)   Individual MCR configuration (read)
        Resource #3)   Existing end-to-end connection configuration (read)
Obligation #2)   To configure multicast routers to support the customer configuration
    Activity #2)   Request MCR configuration from MCR providers
        Resource #4)   MCR configuration (create, read, modify, delete)
Obligation #3)   To configure tunnels between multicast routers to support the customer configuration
    Activity #3)   Request end-to-end connection configurations from network provider
        Resource #5)   End-to-end connection configuration (create, read, modify, delete)
Obligation #3)   To report to the customer service manager any faults or quality of service degradation in the multicast network
    Activity #4)   Map MCR and end-to-end configuration status to multicast network status
        Resource #6)   MCR configuration and status (read)
        Resource #7)   End-to-end connection configuration and status (read)
        Resource #8)   Multicast Network configuration and status (create, read, modify, delete)

Figure 4.33: Example role specifications

4.6.2.6 Management Function Requirements

The management functions required by the identified roles from the service as a whole are derived from the role specifications indicated in the previous section. A portion of the requirements generated for the MMC Customer Service Manager (CSM) is given below:

Register End Users

The MMC CSM must be able to identify a set of end users who have the exclusive right to use the multicast network for multimedia conferencing activities. The CSM must be able to modify this list at any time.

Create Multicast Network

The CSM must be able to specify the extent of the multicast network needed to support the service end users. This is specified by identifying all customer network sites hosting registered service end users and an item of network terminal equipment at each site which will act as an MCR. The CSM must also be able to specify the minimum total quality or service that will be available for multimedia conference traffic transmitted from one end user site to another.

Modify Multicast Network

The CSM must be able to add or remove end user sites and change the network terminal equipment used as the MCR at any time.

Monitor Multicast Network

The CSM must at any point in time be able to monitor whether any end user site can successfully transmit its end users' multimedia conferencing traffic to any other site or set of sites. The CSM must also be able to monitor the minimum quality of service available for these transmissions.

The Provider Service Administrator (PSA) places the following functional requirements on the service:

Build Multicast Network based on Customer Requests

The PSA must be able to map the customer request for a multicast network connecting end user sites to a network consisting of MCRs and tunnels. The PSA must also be able to configure the quality of service configuration of the MCRs and tunnels to meet the customer's minimum requirements for quality of service between end user sites.

Modify Multicast Network

The PSA must be able to reconfigure the network of MCRs and tunnels either in response to the customer's request for changes to the multicast network, in response to the changing status of MCRs or tunnels, or in order to optimise some operational parameters such as cost of resources, reliability, or level of redundancy.

Monitor Multicast Network

The PSA must be able at any time to monitor the status of all MCRs and tunnels either with regard to operational status or delivered quality of service. The PSA must be able to continuously map this status to the multicast network seen by the customer.

4.6.3 Design

4.6.3.1   Information Model

The information model for the MMC service is derived from the basic resource model given in section 4.6.2.3. Descriptions of the individual MOs defined are given in Table 4.5.

Resource	Managed Object Class
An *end user site* is the basic component of the multicast network provided to the MMC service customer. It defines the terminal equipment that the MCR for this site is based on, the public network address of the site, the minimum QoS required for multimedia conference traffic transmitted from this site. It also defines the end users at this site who are registered to use the MMC service.	endUserSite (derived from "X.721:top")
A *multicast network node* represents a single node in the service multicast network. It defines the links (i.e., tunnels) that it has with other multicast network nodes and the QoS available for transmission over these links. It also refers to the MCR that will operate at this node.	multicastNetNode (derived from "X.721:top")
A *multicast router* represents the entity that performs the routing of multicast IP packets between adjacent multicast network nodes and to and from end user site terminal equipment hosting MMC service end users.	multicastRouter (derived from "X.721:top")
A *multicast tunnel* represents the connection between a pair of multicast routers and maps onto an end-to-end network connection.	multicastTunnel (derived from "X.721:top")
A *conference* represents the multiway communication between a set of end users	conference (derived from "X.721:top")

Table 4.5: Information model description for the multimedia conferencing service

The relationship between these managed objects and the distribution of this model between the various management domains involved in the service (corresponding to service stakeholders' domains) is shown in Figure 4.34.

In this model, the end user is represented by a registration number administered by the customer and recorded in a generic *end user* object held in the customer's domain. The network terminal equipment which operates as multicast routers is also represented by a generic managed object held in the domain of the organisation providing the MCR service (which in the case of end user sites is the customer itself). A link between a pair of multicast network nodes is mapped onto a representation of an end-to-end connection in the network provider's domain. The exact representation of this is not within the scope of the MMC service definition but is defined by the management interface to the service offered by the network provider. For instance, in the case of PREPARE where the network provider is a VPN provider (see section 4.3) the end-to-end connection will be represented by a *user stream*.

The *end user site* objects, which represent the customer's request for a multicast network, are each supported by one *multicast network node* which details the network of *multicast tunnels* connecting these multicast network nodes plus any others that are required between end user sites. Each multicast network node is mapped onto a single *multicast router* in the domain of an MCR provider. Each connection between a pair of multicast network nodes is mapped onto an end-to-end connection in the network provider's domain. In the MCR domain each of these connections is represented as a *multicast tunnel* to another multicast router in the same or in a separate MCR provider domain.

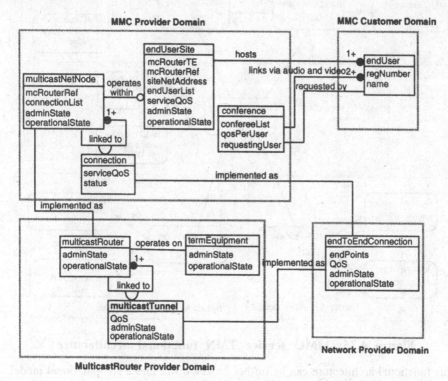

Figure 4.34: MMC service information model relationships and domain distribution

4.6.3.2   Computational Model

The functional architecture for the service can be broken down into domain-specific OSFs and WSFs as shown in Figure 4.35. As mentioned previously, the network provider service is not considered part of the MMC service directly. Its computational model is therefore left unspecified here. This computational model is given in overview in Figure 4.36.

The MMC, MCR and network providers all follow the architecture of having a service OSF from which inter-domain reference points are established and WSFs for service manager and financial agent roles. The MMC Customer has two OSFs. One supports the central management functions, providing access for the Customer Service Manager and Financial Agent role WSFs, while the other, the End User Agent OSF, provides the end user with access to management functionality specifically for requesting resources for conferences. This separation allows for several End User Agent OSFs to be instantiated in customer domains separate from the one containing the central MMC Customer OSF.

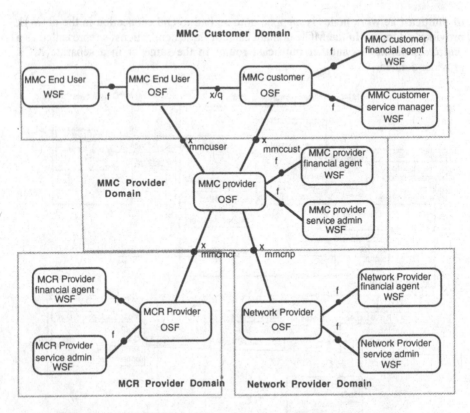

Figure 4.35: MMC service TMN functional architecture

This functional architecture can be further broken down into a computational model that addresses the following areas of functionality:

- *endUserAgent*. Registers conference creation requests from end users with the MMC Provider OSF conferenceMgr.

- *confstatusMgr*. Monitors the status of registered and operational conferences.

- *endUserSiteMgr*. Configures and monitors the customer view of the multicast network as defined by end user site configurations.

- *conferenceMgr*. Monitors the commencement and termination of conferences and logs these events for billing purposes.

- *enduserRegistrationMgr*. Records which end users are registered to use the service at which sites.

- *multicastNetworkMgr*. Receives requests for multicast network configurations from the customer OSF and maps them to end-to-end connections and MCR configuration requests. It also monitors the status of end-to-end connections and MCRs as well as the delivered QoS to the customer, reflecting this status to the provider Service Administrator and the Customer OSF.

- *e2eConnectionMgr*. Receives requests for end-to-end connections from the multicastNetworkMgr and forwards them to the Network Provider OSF.

- *mcRouterMgr.* Manages the configuration of a domain's MCRs in response to requests from the MMC Provider OSF and the local service administrator. It also monitors the status of the MCR and forwards this to the MMC Provider OSF and the local administrator.

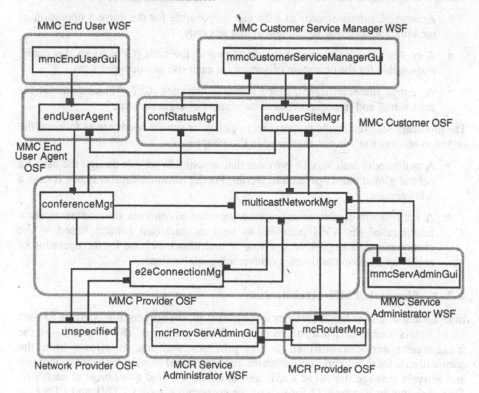

Figure 4.36: Computational model for MMC service functional components

4.7 Application Example

The application example presented here is taken from the phase 2 communications management demonstrator assembled by the PREPARE project. It provides a specific enterprise situation within which the different services described in this chapter can be integrated to meet specific customer requirements. The organisational situation is first assessed by identifying the different real roles involved and the service-specific roles they will undertake. The physical TMN architecture built up to support the various management services over the broadband network testbed is given, together with its mapping to the TMN functional architectures for the various services involved.

4.7.1 Organisational Situation

As discussed in chapter 2, the overall organisational situation chosen is that of a multinational corporation using multimedia teleservices available on the open service market between three of its European sites (London, Copenhagen, and Berlin).

The corporation's management is structured so that individual sites operate their own budgets and may buy and sell services within the corporation. The purchasing of resources from outside the corporation, however, is handled through a central office. The operational infrastructure therefore consists of the following roles:

- A *network administrator* at each site responsible for the correct operation of network and terminal equipment at that site only.

- A *central service manager*, based at one of the sites (CPN #1 in this case), responsible for the operation of services at each site and between sites.

- A *central financial agent*, based at one of the sites (CPN #1), responsible for contractual and financial interactions with other organisations.

The providers selected by the corporation to satisfy its service needs are as follows (the selection process itself is not examined in this chapter):

- A multimedia mail service provider that supplies its service though the use of a central global store operated in Berlin. For its communications needs it uses a VPN service.

- A multimedia conferencing service provider which uses the communication facilities of the VPN provider as well as multicast routers based at the customer's sites to provide a private IP multicast backbone for the operation of multiway desktop multimedia conferencing applications.

4.7.2 Physical TMN Architecture

Independent of any external services it subscribes to, the corporation operates its own set of TMNs, one at each site, based on open TMN platforms. This is used by all the management services outlined in the previous sections. It allows both the corporation's local network administrator and the central service manager to monitor and actively manage the ATM LANs and attached terminal equipment at each site. This is shown in Figure 4.37 for two of the corporation's sites CPN1 and CPN2. In CPN1, the CPN1 Service OS (S_OS) provides management functions to a person acting as the local network administrator, central service manager, and central service financial agent. This person accesses these management functions through the WSF on CPN1 WS. In CPN2, CPN2 S_OS provides management functions for the local network administrator only via the WSF on CPN2 WS. Since the corporation administers each site as a semi-autonomous domain, each site has its own TMN. Therefore the central service manager needs to monitor and manage the operation of CPN2 via an X interface, $X_{cpn1\text{-}cpn2}$ in Figure 4.37.

Also shown in Figure 4.37 are instantiations of the management systems for the service providers which have been described in this chapter. Two public network providers, each with its own TMN, cooperate via the X_{pn} interface. The public network providers provide communication lines to both of the corporate sites and the site housing the multimedia mail provider's global store.

The VPN provider's management services are offered from the VPN service provider's service OS (VPN S_OS). A single operator is able to play the roles of VPN service administrator, VPN service integrator, and VPN financial agent via the VPN provider WSF operating in the VPN WS. The end-to-end public network service offered by the cooperating public network providers is offered as a PuNO service to the VPN provider via the $X_{pn\text{-}vpn}$ interface.

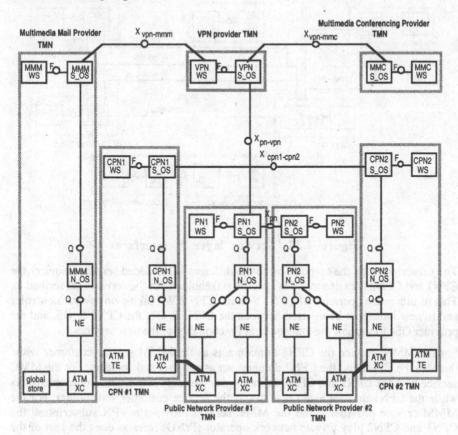

Figure 4.37: Physical TMN architecture for PREPARE second phase demonstrator

The multimedia mail (MMM) management services are offered from the MMM service provider's service OS, MMM S_OS. This OS also accesses, via the $X_{vpn\text{-}mmm}$ interface, the VPN provider's service which is used by the MMM service provider to manage the end-to-end communication services needed for the MMM service. This enables a single person operating the WSFs on the MMM WS to act as the global store administrator, the MMM service administrator, and the MMM service financial agent as well as acting as a VPN customer service manager and a VPN customer financial agent.

A similar situation exists for the multimedia conferencing service provider. This is managed from a service OS, MMC S_OS, which offers management services to the customer. This OS also uses the services of the VPN provider for setting up end-to-end communication links between multicast routers. This allows a single operator using the MMC WS to access WSFs to act as MMC service administrator and MMC service financial agent as well as a VPN customer service manager and a VPN customer financial agent.

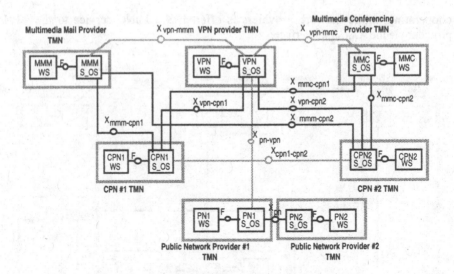

Figure 4.38: Service layer X interfaces

The subscription by the corporation to the different value-added services requires the CPN1 and CPN2 domains to act as specific stakeholders for the services subscribed to. This results in the operators of CPN1 WS and CPN 2 WS taking on several new roles and in new interfaces being used between the CPN1 S_OS, the CPN2 S_OS, and the provider OSs to support the management functionality of the new services.

For the MMM service the CPN1 domain acts as the MMM service customer while the both the CPN1 and the CPN2 domains act as end user stakeholders. For the MMC service both the CPN1 and CPN2 domains play the roles of multicast router providers while the CPN1 domain again operates as the service customer stakeholder. For the MMM service provider's and the MMC service provider's VPN subscription the CPN1 and CPN2 play private network operator (PrNO) roles as does the part of the MMM service domain that manages the global store site.

Based on this specific instantiation of physical TMN components and service instantiations, the tables below summarise the following:

- How the various domains in the scenario map to service stakeholders for the individual services (Table 4.6).

- How the various service layer OSs perform the OSFs of the services specified in sections 4.3 to 4.6 (Table 4.7).

- How the X interfaces required for the final set of service subscriptions (shown in Figure 4.38) support the x reference points identified in the functional architectures for the individual services (Table 4.8). Where an x reference point appears in the functional architecture of more than one service. the one from the service whose stakeholder is the service customer is given in brackets.

Domain	Service Stakeholder Roles Played		
	VPN Service	*MMM Service*	*MMC Service*
PN Provider #1	PuNO	N/A	N/A
PN Provider #2	N/A	N/A	N/A
CPN #1	PrNO	MMM Customer MMM End User	MMC Customer MCR Provider
CPN #2	PrNO	MMM End User	MCR Provider
VPN Provider	VPN Provider	End-to-End Communications Provider	Connection Provider
MMM Provider	VPN Customer	MMM Provider	N/A
MMC Provider	VPN Customer	N/A	MMC Provider

Table 4.6: Domain to stakeholder mapping

Service Layer	Service OSFs		
Operations Systems	*VPN Service*	*MMM Service*	*MMC Service*
PN1 S_OS	PuNO OSF	N/A	N/A
PN2 S_OS	N/A	N/A	N/A
CPN1 S_OS	PrNO Prov. OSF	MMM Cust. OSF MMM End User OSF	MCR Prov. OSF MMC Cust. OSF
CPN2 S_OS	PrNO Prov. OSF	MMM End User OSF	MCR Prov. OSF
VPN S_OS	VPN Prov. OSF	Network Prov. OSF (VPN Prov. OSF)	Network Prov. OSF (VPN Prov. OSF)
MMM S_OS	VPN Cust. OSF PrNO Prov. OSF	MMM Prov. OSF	N/A
MMC S_OS	VPN Cust. OSF	N/A	MMC Prov. OSF

Table 4.7: Service OSF to physical OS mapping

X interfaces	x reference points
$X_{vpn-mmm}$	$x_{vpncust}$ $x_{vpnprno}$
$X_{vpn-mmc}$	$x_{vpncust}$ (x_{mmcpn})
X_{pn-vpn}	x_{custvp} $(x_{vpnpuno})$
X_{pn}	x_{custpn}
$X_{mmm-cpn1}$	$x_{custprno}$
$X_{mmm-cpn2}$	$x_{custprno}$
$X_{vpn-cpn1}$	$x_{vpnprno}$
$X_{vpn-cpn2}$	$x_{vpnprno}$
$X_{mmc-cpn1}$	$x_{custprno}$ $x_{mmccust}$ x_{mmcmcr} $x_{mmcuser}$
$X_{mmc-cpn2}$	$x_{custprno}$ x_{mmcmcr} $x_{mmcuser}$

Table 4.8: X interface to x reference point mappings

4.8   Summary

In this chapter we provided examples of what it means in practice to develop inter-domain management services that are to operate in a multi-domain environment. The salient points that should be noted by the reader are:

- Most of the material in this chapter is directly based on actual PREPARE design document and therefore offers a clearer picture of the inter-domain management service analysis and design process discussed in chapter 3.

- As organisations play multiple parts in a service market, the management systems they use must be able to encompass different functionalities. In a TMN-based system this involves the OSs used by an organisation performing the OSFs corresponding to the stake the organisation holds in each service it is involved in.

- Even quite simple organisational situations with a few services result in a large number of OSFs and reference points needing to be adopted, defined, and/or refined. Mechanisms are needed for handling this increasing complexity.

5 The Broader Context

5.1 Introduction

The results achieved by the PREPARE project reflect the status of the standards available when the project started. In this chapter we provide overviews of what we consider to be the main inter-domain management system development work performed by other organisations in parallel to our work. We concentrate on those developments which we believe should be taken into account, and which we would take into account were we to start all over again with a PREPARE-like project, and we describe our view of their relevance to inter-domain management.

PREPARE took a pragmatic approach to the design of inter-domain management systems based on the TMN framework defined by ITU. Instead of inventing our own, new management system framework we used the existing standards as much as possible. The core of the TMN architecture is the OSI Systems Management standards which explicitly recognise systems management as a distributed application [X.701]. In recent years we have seen a growing interest in distributed systems which has resulted in important new areas of standardisation.

In the area of basic standards and fundamental aspects of telecommunications and (distributed) systems management the following are discussed:

- The ISO's standardisation of *Open Distributed Processing* (ODP) which provides the international standards framework and a reference model for open distributed systems.

- ISO's *Open Distributed Management Architecture* (ODMA) which can be seen as an evolution of the OSI Systems Management standards approach, but which broadens the scope of OSI Systems Management and in this attempt applies the principles prescribed in the context of ODP, and additionally enhances ODP.

- The industry consortium Network Management Forum's (NMF) *Service Management Automation and Re-engineering Team* (SMART) initiative, which represents possibly the first attempt towards the definition of standard service management interactions between providers and their customers.

The growing importance of software-based telecommunications services and their management is reflected in the growing recognition of the need for architectural frameworks for both. Such frameworks aim at facilitating interworking and interoperability, and there is a recognised need for integration or harmonisation of architectural frameworks applied in specific areas (see also Appendix D).

- The industrial TINA Consortium's (TINA-C) *Telecommunications Information Networking Architecture* (TINA) represents an attempt to provide a common conceptual and architectural framework for software for telecommunications services and their management. Based on ODP, Intelligent Network (IN) concepts, TMN, and OSI Systems Management, the approach seems promising with its strength lying in pre-competitive harmonisation.

- Recently, much attention has been paid to emerging new object technologies developed within the industry consortium Object Management Group (OMG).

This group is working on two major areas of an *Object Management Architecture* (OMA) and a *Common Object Request Broker Architecture* (CORBA), which are aspects of general purpose computer application architectures.

For each of these developments we outline its most important characteristics, its scope, and our view on how it relates to inter-domain management.

5.2   The Reference Model of Open Distributed Processing

5.2.1   Background

The rapid growth of distributed processing has led to a need for a coordinating framework for the standardisation of Open Distributed Processing (ODP). The reference model for ODP (RM-ODP), jointly developed by ISO/IEC and ITU in recent years, provides such a framework. It creates an architecture within which distribution, interworking, and portability can be integrated to enable and support interoperability and reusability.

The objective of this standardisation effort is "*the development of standards that allow the benefits of distribution of information processing services to be realised in an environment of heterogeneous IT [Information Technology] resources and multiple organizational domains. This comprises the definition of infrastructure components and functions to accommodate difficulties inherent in the design and programming of distributed systems*"[X.901].

The RM-ODP is specified in four parts providing an overview [X.901], the terminological foundations [X.902], the architecture [X.903], and the architectural semantics [X.904].

5.2.2   Overview

5.2.2.1   Properties of ODP Systems

The RM-ODP identifies several *properties* of ODP systems which must be taken into account in ODP system development and which are addressed in the ODP recommendations [X.901]. For example:

Openness. The property enabling both the contribution of portability (i.e., the ability to execute different components on different processing nodes without modification), and interworking (i.e., the ability to support meaningful interactions between components, possibly residing in different systems) needs to be supported by an ODP system. *Interoperability* should be supported by an integrated use of a set of well established standardised communication systems, covering interceptors (or proxy systems) for protocol mappings.

Modularity. The ODP system design should be modular, i.e., components of the ODP system should be designed as autonomous building blocks interacting with other components by adequate interaction mechanisms (e.g., standardised communication protocols).

Integration. An ODP system should allow for the integration of different technologies and architectures (communication systems and data repository systems) to overcome heterogeneity problems of the different standards worlds[1]. In an inter-domain management context, this should for instance cover the integration of SNMP[2] [SNMP] based systems in customer domains [NMF-TR107] [NMF-028] and the CORBA technology (see section 5.6).

5.2.2.2   Architecture

The ODP Reference Model defines an *architecture* which contains the specification of the required characteristics that qualify distributed processing systems as open. These are the constraints to which ODP standards must conform. Part 3 defines the ODP framework as comprising the *five viewpoints*, the *viewpoint languages*, the specification of *ODP functions*, and *transparency prescriptions* showing how to use the ODP functions in order to achieve distribution transparencies [X.903].

The complete specification of large ODP systems is a very complex non-trivial activity which requires a structured approach. In order to be able to handle the complexity of the whole system, the ODP viewpoints allow for the separation of different concerns of an ODP system. The ODP viewpoints and their languages provide a set of concepts and structuring rules for specific concerns.

The *enterprise, information, computational, engineering*, and *technology* viewpoints have been chosen as a necessary and sufficient set to meet the needs of ODP standards. Viewpoints can be applied, at an appropriate level of abstraction, to a complete ODP system, in which case the environment defines the context in which the ODP system operates. Viewpoints can also be applied to individual components of an ODP system, in which case the component's environment will include some abstraction of both the system's environment and other system components. The viewpoints are characterised as follows: the *enterprise viewpoint* focuses on the purpose, scope, and policies of a system; the *information viewpoint* focuses on the semantics of information and information processing; the *computational viewpoint* enables the distribution of functionalities through the decomposition into objects which interact at interfaces; the *engineering viewpoint* focuses on mechanisms and functions required to support distributed interaction between objects in the system; the *technology viewpoint* focuses on the choice of technology in that system.

Another concept which needs to be considered for ODP systems is *distribution transparency* [X.903]. Distribution transparency is the property of masking from applications the details and the differences in mechanisms used to overcome problems caused by distribution. This is a central requirement that arises from the need to facilitate the construction of distributed applications. Aspects of distribution which should be masked (totally or partially) include: heterogeneity of supporting software and hardware; location and mobility of components; mechanisms to achieve the required level for QoS in the case of failures (e.g., replication, migration, checkpointing, etc.). A number of specific distribution transparencies are identified in

---

[1]   As promoted by standards bodies such as the Internet Engineering Task Force, ISO, ITU, and OMG.

[2]   SNMP (Simple Network Management Protocol) is the Internet management protocol. It is used to manage internets and is in widespread use for the management of small networks, such as local area networks.

Part 3, for example: *access transparency* which masks differences in data representation and invocation mechanisms to enable interworking between objects; *location transparency* which masks the use of information about location when identifying and binding to interfaces; *transaction transparency* which masks the coordination of activities amongst a configuration of objects to achieve consistency.

To assist the sound and uniform development of formal descriptions of ODP-based standards and of ODP systems a set of specification languages may be used (LOTOS, SDL, ESTELLE or Z). Architectural semantics are therefore required for ODP to provide a formalisation of the ODP modelling concepts [X.904].

5.2.3   Relevance to Inter-Domain Management

The experience of PREPARE has shown that the development of open inter-domain management systems is a difficult and complicated matter because of the complexity of the problem being considered. The RM-ODP provides, with its set of viewpoints, a *standardised* structuring principle which is much needed, especially for a multi-organisational development approach. The basic properties identified for ODP systems are equally important for open inter-domain management systems (e.g., openness and integration). Therefore, the RM-ODP seems to be highly relevant as a standardised approach to the development of open inter-domain management systems.

The RM-ODP, with its viewpoint languages and its consistency rules between the viewpoints, provides an adequate framework enabling a comprehensive iterative system design. It is expected that such a development approach will lead more efficiently and more quickly to a suitable inter-domain management system. This will become important in an open service market where service providers are in competition with other service providers.

However, the use of the RM-ODP to develop open inter-domain management systems faces two important difficulties. First, the RM-ODP provides some broadly defined principles for ODP systems and does not provide management-specific extensions. Second, it requires a non-trivial amount of effort by developers before they can use the principles "fluently". The first problem is already being approached by standards bodies in the context of the *Open Distributed Management Architecture* (see the following section).

The RM-ODP provides a set of further essential principles and concepts which are required to achieve inter-domain management on the global scale. It would be beneficial to adopt these standardised concepts for the design of inter-domain management systems. Otherwise similar concepts would have to be developed. However, even though the RM-ODP would seem to be necessary, it is not specific enough to provide all the concepts and guidelines required for a comprehensive design of inter-domain management systems. It was developed as a general architectural framework for open distributed processing systems. This architectural framework must be refined for the problem of distributed management in the open service market and guidelines must be developed for its use in the telecommunications world[1].

---

[1]   Section 5.5 briefly introduces the TINA initiative as a response to this issue.

5.3   Open Distributed Management Architecture

5.3.1   Background

In 1994 ISO/IEC started a new project on *Open Distributed Management* (ODM) in order to provide a generic concept for an *Open Distributed Management Architecture* (ODMA) and an ODP-based model of OSI Systems Management [ODMA].

Although the current scope of this activity is focused on modelling distributed management, it is planned to start further activities on the management of distributed applications in the future. This migration from OSI Systems Management towards the management of distributed applications will lead to an integration of ODM-related questions as an integral part of the ODP standards effort.

5.3.2   Overview

ODMA provides an architecture for the specification and development both of systems management as an open distributed application and of the management of open distributed applications. The management will be of a distributed nature. This implies:

- Distribution of the managing activity.
- Management of distributed applications.
- Management of resources that may be distributed.

ODMA is supplementary to OSI Systems Management [X.701]. It describes how the OSI Systems Management architecture can be reused in a distributed environment. ODMA embeds all aspects of OSI Systems Management. In the limiting case of interaction between a single system carrying out management activity and another single system where the resources being managed are located, OSI Systems Management as defined in X.701 may apply. When applications or resources to be managed or the managing applications themselves are distributed, the ODMA architecture should be used.

ODMA is compliant with the RM-ODP (see section 5.2) so that in a distributed environment OSI Systems Management can be used in combination with other techniques that are engineered and implemented according to ODP principles. The specification of other techniques for communicating management information is outside the scope of ODMA [ODMA].

The ODMA defines a general framework for open distributed management. It abstracts from specific interpretations of management such as the OSI Systems Management and the ODP interface definition language (IDL) paradigm. However, within this framework these specific interpretations will be elaborated. ODMA should evolve gracefully from its current model to a fully distributed model. This evolutionary path should take into account some limitations that are present in current systems and not preclude evolution to full RM-ODP compliance. To allow further interpretations in the future, the ODMA committee draft currently consists of three parts, but allows for the development of other parts.

Part 1: General Framework. This part describes ODMA as a specific interpretation of the RM-ODP for the purpose of managing distributed resources, systems, and applications. ODMA focuses on the specific features or requirements of management which are not already reflected in the RM-ODP. It describes an interpretation of the five viewpoints for open distributed management and identifies access and location transparency (see section 5.2) as the most important distribution transparencies for ODMA, at least in the first stage. It introduces general terms that are needed for open distributed management. It may also identify tools for describing the open distributed management applications.

Part 2: OSI management interpretation of ODMA. This part describes the OSI Systems Management interpretation of ODMA. It relates current OSI Systems Management concepts to ODP and ODMA concepts. However, it also extends the interpretation of the current OSI Systems Management standards. Since this is a specific interpretation, limitations may be imposed on the general principles of ODMA. For instance, only a number of distribution transparencies may be supported by the OSI management mechanisms.

The operational interface specification requires the representation of message exchanges between objects. These exchanges are indicated by the specifications defined in GDMO [X.722]. The GDMO templates of actions and notifications will be visible from the computational viewpoint together with the operation qualifiers on attributes (e.g., set, get). The behaviour of an object as a consequence of the receipt of an operation is specified in the information viewpoint. Since OSI Systems Management does not specify where scoping and filtering mechanisms are located (i.e., in the application acting in the role of manager or in the agent) interoperability problems may occur. As an enhancement to OSI Systems Management, the precise definition of operational interfaces covering all parameters will solve this potential interoperability problem.

In OSI Systems Management location transparency can be achieved by global naming, i.e., managed objects can have a globally unique name independent of the systems that provide access to those objects (i.e., agents) (see Appendix B). OSI Systems Management managed objects are named according to a naming schema using name binding templates as specified in GDMO. In ODMA a global naming tree will be the most practical solution for managed objects. The selection of objects using scoping and filtering needs to be applicable[1] within such a tree. ODMA can support scoping and filtering in combination with the directory [X.500]. The directory should store the management information tree or should store the references to the managed objects according to the naming schema.

Part 3: "OMG, CORBA, IDL and ODP supporting functions": support of ODMA. This part describes OMG, CORBA, IDL and ODP supporting functions and is for further study.

5.3.3   Relevance to Inter-Domain Management

As already stated in section 5.2, the use of ODP principles for inter-domain management is important to enable an adequate modelling of TMN interworking and the development of distributed management systems. ODMA will provide an ODP-based architecture for open distributed management and specifically an extended interpretation of OSI Systems Management. As described for the examples of scoping,

---

[1]   Scoping identifies the objects to which filtering is applied, e.g., by identifying a root of a subtree. Filtering is used to select a subset of the objects identified by scoping (see Appendix B).

filtering, and global naming, the OSI Systems Management concepts have been enhanced to increase interoperability between remote systems. Organisations and projects developing distributed management systems need to be aware of this standardisation work[1].

ODMA could be seen as a necessary starting point providing a basic architecture which needs to be enhanced to cover the scope of the TMN architecture and management of distributed applications (such as multimedia teleservices or distributed end-to-end management services). It is expected that new work items on distributed management will be established when the basic architecture (ODMA Part 1) and the interpretation of OSI Systems Management (ODMA Part 2) becomes stable.

In relation to TMN, OSI Systems Management is well-established as the communications and information modelling approach, both for intra-domain management and, as presented in this book, for inter-domain management. As TMN evolves towards adopting more open distributed processing aspects, ODMA appears to be a natural companion standards area, providing enabling technologies just as OSI Systems Management does for TMN as it currently stands.

5.4   Network Management Forum

5.4.1   Background

The Network Management Forum (NMF) is an industrial consortium for various types of organisation (e.g., public network operator, equipment vendor, customer) to agree on industry standards for interoperable network management systems. The NMF's work is divided into three work programmes: *OMNIPoint*, *Service Provider Integrated Requirements for Information Technologies* (SPIRIT) and *Service Management Automation and Re-engineering Team* (SMART). In the following we briefly describe these and in particular we consider the relevance of SMART for inter-domain (service) management.

5.4.2   Overview

OMNI*Point* aims to enable the automation of end-to-end management by providing detailed specifications that match TMN standards and can be applied to business problems. In the *OMNIPoint Strategic Framework* it is stated that *"OMNIPoint is a comprehensive framework for managing networks and systems from the perspective of the services they deliver to end customers. The aim of OMNIPoint is to make advanced, integrated service management happen across the industry as fast as possible, at least cost, and at least risk."* [NMF-SF].

OMNI*Point* includes a *Management Systems Framework* (MSF) which aims to provide an integrated view on what a management system is:

"The OMNIPoint Management Systems Framework (MSF) addresses the integration of two frameworks necessary to achieve fully-automated service management:

---

[1]   Concepts elaborated in PREPARE for a distributed TMN architecture and an inter-domain management infrastructure, such as the use of the X.500 directory, have been adopted by ODMA for global object naming and the support of location transparency.

- *an Object-Oriented Framework that enables data and function abstraction as well as application distribution, and*

- *a Management-Agent Framework that addresses the relationships and communications between manager systems, agent systems and managed resources."* [NMF-SF].

The MSF comprises various components: communications protocols, access facilities, common services, various types of object (adapter, composite, intelligent), and applications. In outlining the roadmap in terms of OMNI*Points*, the NMF starts from a set of defined functional elements and a set of infrastructure elements and explains other OMNI*Points* in terms of additions to the initial sets. Currently, OMNI*Point* 1 and OMNI*Point* 2 have been issued. The MSF is further elaborated in the *OMNIPoint Integration Architecture* [NMF-TR114] which identifies engineering solutions, such as CORBA (see section 5.6).

The SPIRIT group within the NMF is concerned with defining a common set of software specifications for a general purpose computing platform for the telecommunications industry. The term *general purpose* means that the platform is not aligned with a specific application type (but still provides a suitable platform for management application development). Generally desired qualities of the general purpose computing platform are the following: portability, interoperability, and modularity. The stated benefits of a platform cooperatively specified by several actor types in the telecommunications market are multivendor software environment, integrated systems, and technology independence [SPIRIT].

5.4.3 · SMART

In 1994 the NMF identified a need to determine what service providers require in the way of common industry interfaces to support automation. A team called Requirements Capture (RECAP) was formed and began interviewing service providers. However, it became clear that all service providers involved had to have a common way of viewing service management processes in order to agree on specific interfaces. A business process model (BPM) was defined for that purpose [NMF-BPM]. The RECAP project ended and a number of interface requirement-specific projects within the NMF began in a work group called SMART.

In SMART, the emphasis is on the flow of information between corporate entities: from end customers to service providers, between service providers, and between service providers and suppliers. Every service provider plays two roles: provider towards its customers, and customer towards other (sub-)providers.

The BPM identifies a set of generic business processes taking place within service provider organisations. These form the basis for SMART's information requirements definition and protocol specification work. The business processes so far addressed by SMART include ordering, performance reporting, problem management, and billing.

5.4.3.1   Ordering

The objective of the SMART ordering team activity is to define and deliver an automated and standardised order submission, tracking, and confirmation interface for the use of telecommunications service providers. The interface in question is that between a main service provider and a subcontracted service provider [SMART-Ord].

The scope of the SMART ordering process covers submission, tracking, and confirmation of the delivery of a service ordered from an existing service portfolio across the service provider to service provider interface. It covers enquiry (pre-service), new service orders, changes to existing services, and cancellation of services. The information model but not the communication model is covered.

An end-to-end ordering process model and an information model have been produced which has led to the identification of a list of variables that could be tracked across the ordering interface.

5.4.3.2   Problem Management

The objective of the problem management activity is to define and develop an open interoperable interface for problem management between service providers and customers. This interface has been given the provisional name of the P interface [SMART-P-IF].

The P interface defines the trouble management functionalities between a customer and a service provider. It provides an interoperable interface by means of which the customer can report troubles, track, and manage trouble reports on resources or services offered by the service provider. In supporting the customer's view of trouble management, it seeks to provide increased quality of service by ensuring that troubles reported by the customer receive attention and are cleared to restore the service to its previous level of capability.

The P interface defines a uniform trouble management interface between the customer and the different service providers, despite differences between the service providers. The customer no longer has to enter trouble reports using an interface specific to one service provider.

5.4.4   Relevance to Inter-Domain Management

Some distinguishing characteristics of the SMART work in the NMF makes it particularly interesting for the types of problem addressed by PREPARE. The Business Process Model (BPM) upon which the specification work is based provides an alternative, process-oriented view on a service management organisation [NMF-BPM]. This is a useful input for structuring the diverse functional requirements. The BPM provides an approach which enables the scope of a specification to be defined more clearly in terms of:

- Functional scope by separating the tasks at hand from, and in the context of, the general and broadly defined service management task.

- Organisational scope (intra-domain and inter-domain) by allowing functional (idealised) tasks to be related to notional agencies within an organisation and, quite importantly, between organisations.

- It offers a *situated view* of a particular management task by relating it to other (or all) processes (management tasks) of a stakeholder organisation (service provider and customer in the case of SMART).

The most important aspect of SMART is, however, that it represents a first attempt to define generic service management information models. As such, the expected information specifications from SMART are highly relevant to the inter-domain management problem. These specifications have the characteristic that no specific management communications protocol is assumed or prescribed. In fact, it is a requirement that the (information) specification is generic enough to be specialised to a number of communications mechanisms (such as CMIP, SNMP, and Remote Procedure Call).

For the protocol independence property it still remains to be seen what effort is required to transform such a generic information specification into specialised technical specifications for specific protocols for actual interoperable interfaces.

5.5   Telecommunications Information Networking Architecture

5.5.1   Background

The Telecommunications Information Networking Architecture Consortium (TINA-C) is an international industrial consortium consisting of network operators and telecommunications and computer equipment vendors. This consortium is aiming to develop an open architecture for telecommunications services in a multi-provider, multi-vendor environment [TINA-018]. The approach is broad, combining service provisioning principles from IN and telecommunications management principles from TMN with ODP concepts to address service provisioning and operation with the management of services and related networks and systems. This approach also includes defining a model for a distributed processing environment for supporting the functional components defined within the architecture.

The overall aim of this work is to provide an architecture for a wide range of telecommunications services from traditional voice-based services to multimedia and information services, together with the accompanying management and operations services, all operating over a variety of technologies. The resulting software architecture must provide a framework for flexible, reusable components in order to ease the construction, testing, deployment, and operation of services as well as hiding the complexity of distribution and heterogeneity of underlying technologies from the service designer. The idea is that of a future service information market where services are offered for purchase to customers on an equal basis regardless of the underlying technology on which the services will run. As well as taking results from the TMN and IN, the consortium works closely with other bodies working in this area, e.g., the Network Management Forum, the ATM Forum, EURESCOM, the RACE programme, and the OMG.

This section gives an overview of the salient points of the TINA-C approach and architecture, its structure for defining and grouping service components as well as examining its relevance to inter-domain management.

5.5.2   Overview

The TINA-C overall architecture can be refined into four technical areas on the basis of the overall architectural principles:

- The *Computing Architecture* provides a set of rules and concepts for how interacting software components (computational objects) are specified and how they communicate.

- The *Service Architecture* provides a set of concepts and principles for analysing, specifying, reusing, designing, and operating service related communications software components.

- The *Network Architecture* provides a set of generic components for using and managing network components.

- The *Management Architecture* provides a set of generic management principles and concepts that are applied to the management of the other architectures.

The following sections provide more detail about the different architectures. In addition, the TINA concept of a domain is briefly described.

5.5.2.1   Computing Architecture

Distributed Processing Environment. The computing architecture specifies that TINA-C based systems adhere to a strict object-oriented paradigm. Therefore, all entities that make TINA compliant systems are constructed from computational objects. A TINA object can have multiple interfaces that are defined as either operational interfaces, which can invoke operations offered by other objects, or stream interfaces, which can exchange stream flows with other objects. These computational objects need to communicate without regard to their location, i.e., distribution transparency. For this purpose an abstract distributed processing model, the Distributed Processing Environment (DPE), has been defined [TINA-005]. The DPE architecture describes at an engineering level the installation, execution, and interactions between distributed applications. The DPE includes additional services such as trader, name, notification, and security services.

Viewpoints. TINA-C advocates the use of the RM-ODP for addressing the development of components that are to comply with the architecture. These viewpoints, as used in the TINA context, are interpretations of the RM-ODP viewpoints described in section 5.2.

Universal Service Component Model. Any TINA service is defined as a set of interacting software components structured in accordance with the Universal Service Component Model. These components are the aspects of a service that must be considered when modelling the service and comprise the following parts:

- A *core sector* that defines the nature of the service and performs the application.

- A *substance sector* by which the core can access external functionality.

- A *usage sector* by which the core offers its own functionality to other service components.

- A *management sector* which offers specific management access.

By partitioning the service component in this way, the core sector can be developed in an independent and reusable way with technological dependencies due to communication requirements with other services being handled by the other sectors. This also allows procedural or functional models, such as IN, to be encapsulated as the core sector. The model can be applied recursively so that each sector can in turn be constructed from other service components adhering to the model.

5.5.2.2   Network Architecture

The TINA-C network architecture aims to describe the transport network used by
services in a technology independent way. A Network Resource Information Model
has been defined which is an information specification of transmission and switch
technologies in which the technology dependencies have been extracted [Lengdell].
This model describes network connectivity as connection graphs made up of ports
connected by lines representing connectivity. The model includes two basic types of
connection graph: a physical connection graph representing physical nodes and the
physical lines connecting them, and a logical connection graph representing the end
points and interconnecting lines of streams that are requested by service software and
mapped to a physical connection graph to set up the actual connections required.

5.5.2.3   Service Architecture

The TINA-C concept of a service is used to cover telecommunications services,
management services, and end user services [TINA-012]. The information viewpoint
of the service architecture addresses the separation of call and connection control and
the separation of user and terminal issues. The former separation eases the separation
of the implementation of service policy, including user involvement, from the
allocation of network resources to set up connections. The latter provides support for
the mobility of end users and terminals. The computational model describes the
structure of a distributed service identifying the following fundamental components:

- *User Agent* establishes the service on the user's behalf and maintains user
  profile and personal service customisation information.

- *Terminal Agent* represents a technology-independent view of the terminal
  attached to the network.

- *Service Session* maintains the state of the service and the involvement of users,
  and constructs the logical connection graphs needed for the service.

- *Subscription Manager* manages customer and user subscription details, and is
  responsible for user access control.

- *Communication Session Manager* receives logical connection graphs from the
  Service Session and sets up the corresponding network connections.

5.5.2.4   Management Architecture

The TINA-C management architecture has a different status from the other
architectures in that it influences all the other architectures [TINA-010]. As there are
different kinds of management it is split primarily into two parts, computing
management and telecommunications management.

Computing management is further broken down into:

- *Software Management*. The deployment, configuration, instantiation,
  activation, deactivation, and withdrawal of software components.

- *DPE Management*. Management of computational objects and their grouping.

- *Computing Environment Management*. The management of the computing
  system supporting the DPE.

- *Kernel Transport Network Management.* The management of the network that supports the communication between DPE nodes.

Telecommunications management is broken down along similar lines to the logical layered architecture in TMN. Service management is related to the service architecture while network management and network element management are related to the network architecture.

The management architecture also addresses different management functional areas using the same categories as the OSI Systems Management framework, namely fault management, configuration management, accounting management, performance management, and security management. For the service architecture the emphasis has been placed on configuration management and, in particular, on the management of sessions. For the network architecture, configuration management has been further subdivided into resource configuration and connection management.

5.5.2.5   Domains

The TINA Management Architecture addresses inter-domain management primarily through the TINA-C domain concept. This aims to partition the management environment into management domains that represent the extent of resources that are subject to the same type of management functionality, e.g., fault management, security management, etc. In addition, administrative domains have been defined that represent the extent of resource ownership by a stakeholder. Although management domains can overlap, administrative domains cannot. Furthermore, a management domain is defined as being wholly contained in only one administrative domain.

TINA also defines how to identify and characterise reference points between administrative domains [TINA-020]. A domain that has a relation with another domain has a role relative to that relation. The relationship between two domains can therefore be defined by the interfaces offered by the roles taken by each of the two related domains. Relationships between administrative domains are further characterised as being either client-server relationships or federal relationships. Based on the layered view of TINA, which is divided into a service layer, a network resource layer, and a DPE layer, ten basic reference points are defined for permitted client-server and federal relations within the same layer and client-server relations between different layers. These basic reference points can be further specialised to suit the definition of inter-domain relations in specific circumstances. This method of defining inter-domain relationships between stakeholders is similar to that used in PREPARE, i.e., defining inter-domain management services via relationships between stakeholders, but with the extensions to accommodate the TINA layered view.

5.5.3   Relevance to Inter-Domain Management

The TINA-C architecture is very broad, covering aspects of service subsuming the issues of inter-domain management discussed in this book. By taking input from many of the existing bodies of work (e.g., NMF, OSI Systems Management, TMN, IN) it provides an architecture that reduces the risks of migrating between different sets of standards by preserving many of their advantages. This, therefore, makes TINA compliance an attractive aim for systems addressing areas of service management, including that of inter-domain management. Furthermore, the division of the TINA architecture and the focus on reusable components offers a good basis for the definition of management services in a modular way.

Through its domain concept, TINA-C would seem to have addressed inter-domain issues at an architectural level. However, the maturity of this work is such that it is not clear how these architectural aspects will be realised in practice. In particular, future implementation of the TINA DPE will have to ensure that security services and suitable transparencies are available before inter-domain management systems developed following the TINA model will be realistically implementable. Currently these capabilities are not available in a TINA DPE implementation. There appears to be little information on the implications of following the TINA architecture in practice and how these system can interoperate with systems already implemented in compliance with existing standards, e.g., IN and TMN.

Like the NMF's SPIRIT, the definition of a common platform specification (DPE) can be expected to result in the possibility of software portability across organisational boundaries and in opportunities for independent software vendors. Besides the consequences for the market and pricing of telecommunications software, a common universal platform would offer the potential of vendor independence at a more finely grained level than aimed for with TMN, through smaller pieces of open interoperable software modules.

5.6   Object Management Architecture

5.6.1   Background

OMG is an international consortium supported by information systems vendors, software developers, and users. OMG was formed to help reduce complexity, lower costs, and hasten the introduction of new software applications. The overall goal of OMG is to provide a common architectural framework for object-oriented applications based on widely available interface specifications. The adoption process is around 16 months and only available software solutions are considered in this process.

The technical goal of OMG is to foster interoperability and portability for application integration through cooperative creation and promulgation of object-oriented specifications based on commercially available software:

- Single terminology for object-orientation.
- Common abstract framework.
- Common interfaces and protocols.

It is believed that conformance to these specifications will make it possible to develop a heterogeneous application environment across all major hardware platforms and operating systems.

The rest of this section provides an overview of the architectural framework, Object Management Architecture (OMA), issued by OMG. Finally, the relevance to inter-domain management is described.

5.6.2   Overview

The OMA defines a reference model and an abstract object model [OMA].

5.6.2.1   OMA Reference Model

The reference model partitions the OMG problem space into practical high-level architectural components. It identifies and characterises the components, interfaces, and protocols that compose OMA but does not define them in detail. The reference model has two system-oriented components: Object Request Broker (ORB) and Object Services, and two application solution specific components: Common Facilities and Application Objects.

The *ORB* is the communications heart of the architecture. It provides the mechanisms by which objects transparently submit requests and receive responses. As stated in the *Object Management Architecture Guide*: *"The ORB provides interoperability between applications on different machines in heterogeneous distributed environments and seamlessly interconnects multiple object systems"* [OMA]. The ORB provides access and location transparency (cf. section 5.2).

The *Common Object Request Broker Architecture* (CORBA) specifications define how to build and use OMA-conformant ORBs [CORBA]. CORBA also includes an ORB interoperability specification, and an initialisation service - a portable way for ORB clients to start up services required by applications. One of the most important features in the CORBA specifications is IDL. It is used by applications to specify the various interfaces they intend to offer to other applications and includes exception handling. Language mappings for C, C++, and Smalltalk are part of the CORBA specifications and additional mappings are expected for almost every commonly used programming language.

Object Services provide the basic operations for the logical modelling and physical storage of objects. Object Services connect directly to CORBA-conformant ORBs and are also supported on all ORB platforms. The services are considered to be common to nearly all object-based applications. These services standardise the life-cycle management of objects and provide a generic environment in which single objects can perform their tasks.

CORBAservices is the collection of *Common Object Service Specifications* which contain the specification of the following object services: Life Cycle, Naming, Event Handling, Persistent Object Service, Relationships, Transaction, Concurrency, and Externalisation [COSS]. A Trading service has also been specified. A number of other services are in the process of being specified.

Common Facilities (or *CORBAfacilities*) comprise higher level services that are useful in many application domains or they are high value capabilities for specific domains or vertical markets. They are not required to be supported by all ORB platforms. The facilities can be divided into four groups: User Interface (e.g., Compound Documents), Information Management (e.g., Compound Storage), System Management, and Task Management (e.g., active agents). The first service expected to be specified is the compound document facility.

The *Application Objects* component represents those applications performing particular tasks for a user. This component is not part of the OMG standardisation activity.

5.6.2.2   OMA Object Model

The Abstract Object Model defines common object semantics for specifying the externally visible characteristics of objects in a standard and implementation-independent way [OMA]. The goal was to define an object model that facilitates portability of applications and type libraries, and interoperability of software components in a distributed heterogeneous environment. The *Core Object Model* is based on a small number of basic concepts: objects, interfaces, operations, and types and subtyping. OMG does not interpret these concepts in any unusual way. Extensions to the Core Object Model can be defined using *Components* which are not required to be supported by all systems. *Profiles* is a mechanism for technology domains to group pertinent components. The Object Model is neither a strict superset nor a least common denominator of existing object models or features from object-oriented programming languages or object-oriented database management systems, but defined in such a way that it is believed to be sufficiently rich to be useful.

5.6.3   Relevance to Inter-Domain Management

Inter-domain management is not directly addressed in any of the OMG specifications. However, the specifications have many (low-level) elements which could support the specification and realisation of inter-domain management, although a fundamental issue like security has been addressed but has not yet been specified. In addition, many of the current ORB implementations do provide access control services. The use of OMG specifications in TMN applications will most probably be in the form of ORBs. CORBA has become an accepted middleware de facto standard and is already available on top of most of the common operating systems and will soon become part of many of these operating systems. Thus, it is most likely that CORBA at least will be part of many TMN OS implementations for internal communication.

In TMN the basic distribution of management systems is at the TMN building block level. Distribution of the functional components is beyond the scope of TMN. A straightforward approach would be to keep the TMN interfaces but to use CORBA as the distribution architecture within the realisation of functional components such as the Management Application Function and Presentation Function (see Appendix A). Hence, communication between objects across building blocks would require conversion of IDL to GDMO and vice versa (the issue has already been addressed by the NMF and X/Open [XoJIDM]), i.e., inter-domain management aspects are expressed in terms of TMN interfaces but internally implemented using CORBA.

Even if there are no signs that inter-domain management in the short or medium term will be exclusively and directly based on OMA/CORBA, in the long term it may be a very realistic implementation option. The possible large-scale penetration of CORBA as the distribution mechanism in the TMN OS implementations could bring about a need for a software architecture for TMN applications. The OMG framework could have then evolved to the point where it would be a viable solution. A common platform specification and software architecture would enable inter-domain management specification and implementation at a more finely grained (management application software) level than currently envisaged with TMN. CORBA is a candidate for the realisation of a support environment for the computational viewpoint in ODP.

TINA-C has recognised CORBA as being a candidate for its DPE[1] (see also section 5.5.3).

5.7   Conclusions

The aim of this chapter was to direct the reader's attention towards the most important directions for new developments having a relationship to inter-domain management. These new developments are not seen here as threats or candidates to replace TMN as the basis for inter-domain management but rather as complementary to the current TMN recommendations.

It should be clear that the initiatives described address complementary aspects which together provide a broad solution to the problem of inter-domain management. As the scope of each development is more or less unique, it is not unlikely that a combination of different options should or could be utilised to solve a particular problem. Assessing the importance and likely impact of these initiatives for the future development of inter-domain management systems can be summarised as follows.

5.7.1   Inter-Domain Management = Open Distributed Management

The integrated broadband communications environment constitutes a globally distributed enterprise characterised by federative organisational structures of multiple cooperating and competing autonomous service provider and customer domains. A set of general open distributed system characteristics needs to be considered for inter-domain management in this environment. The RM-ODP addresses these general characteristics.

In this global environment, TMN applications will constitute very complex, large-scale distributed applications that will run across several different domains, with various enterprises designing and specifying these applications. The cooperative management of telecommunications systems in this environment specifically requires open distributed processing mechanisms.

Centrally controlled TMN systems cannot be applied in such an environment. The RM-ODP provides concepts for open distributed processing systems which can be applied for the design of distributed TMNs supporting such federated enterprises. Inversely, distributed systems must be managed. While the RM-ODP contains basic elements for ODP systems management, a full set of ODP system management functions still needs to be specified.

The development of the ODP-based ODMA will provide solutions both for the management of ODP systems and for a subset of the whole inter-domain management problem. These concepts need to be rich enough to provide solutions for the comprehensive modelling of distributed telecommunications systems and of information technology-based cooperation (or inter-domain management) between multiple autonomous organisations.

---

[1]   At the same time it has been pointed out that the TINA architecture compared to CORBA provides a richer framework with regards to the object model: TINA objects may support multiple interfaces and several objects can be encapsulated in an aggregate construct (Building Block) with its own interface(s) [Kitson].

5.7.2   Common Platforms and Software Architectures

TINA is an obvious long term goal for telecommunications service providers, centred around defining a software architecture for future software-based telecommunications systems. TINA utilises the RM-ODP to structure the architectural aspects and specialises the viewpoint definitions and specification languages to the particular application in telecommunications software development. In the long term, TINA assumes a common universal service infrastructure, the DPE. CORBA has gained a lot of attention as an intermediate implementation option for the DPE although the object model is not as powerful as TINA's [Kitson].

CORBA has also achieved acceptance from the NMF. The NMF has outlined an integration architecture [NMF-TR114] in which CORBA IDL is recommended as the interface definition notation for management solutions in a multi-technology environment (where TMN and SNMP are seen as technologies). CORBA is also part of the NMF's platform procurement specification SPIRIT [SPIRIT].

Realising that several standardised approaches to telecommunications management will coexist (TMN/OSI Systems Management, SNMP, and CORBA) and that each prescribes its own interface definition notation (GDMO, SNMP, IDL) the NMF and X/Open have jointly established a task force to specify translation algorithms (specification translation and interaction translation) between these specification languages (the X/Open NMF Joint Inter-Domain Management[1] working group) [XoJIDM]. In addition, the NMF has specified translation between GDMO and SNMP [NMF-TR107] [NMF-028].

5.7.3   Service Management

Whereas TINA to some extent provides a telecommunications service-specific application of the RM-ODP, especially in the area of requirements definition, it has not addressed in detail the ODP enterprise viewpoint and specification language. SMART, on the other hand, has taken an interesting approach to this topic. The SMART work programme is centred around automating service providers' business processes, especially for service management information exchange between providers. The SMART team has defined a generic model of service providers' business processes which is the starting point for their detailed service management information specifications to be published as part of future OMNI*Point* specifications.

The NMF provides several complementary specifications, some of which can be seen as "populating" TMN in specific areas. Already, some NMF specifications have been adopted by the ITU[2] and the NMF has a formal liaison with ITU. The specifications expected from the SMART work group are likely to provide the basis for short to medium term standardised inter-domain service management automation as they result in the specification of technical solutions.

---

[1]   Note that the XoJIDM concept of inter-domain management refers to the TMN/OSI systems management, SNMP, and CORBA standards domains, whereas PREPARE's inter-domain management concept refers to management administrative domains.

[2]   For example, M.3020 has been revised from the 1992 version to the 1996 version towards a more NMF Ensemble oriented approach [M.3020/94], and the X.790 recommendation for trouble ticket exchange is also based on NMF work [X.790].

6 Conclusions

Since the PREPARE project started its work in 1992, inter-domain management has become increasingly important for supporting the requirements of network operators, service providers, customers, and end users in an open service market.

In the previous chapters of this book, we described the initial goals of the RACE II project PREPARE and the approach taken towards realising two communications management demonstrators. The development methodology devised in PREPARE for multi-domain service management and a number of service management examples have been outlined. In addition, the most important standards work taking place in parallel to PREPARE with respect to inter-domain management was briefly reviewed and its impact on inter-domain management was discussed.

A final assessment of PREPARE and some important areas requiring further work are summarised in this chapter. Finally, an overview of the evolution of telecommunications services and their management is provided.

6.1 Final Assessment

This section presents some final thoughts on the increased significance and impact of inter-domain management and the application of ITU's TMN framework for inter-domain management. In addition, a summary of the implementation experiences from the realisation of the communications management demonstrators is provided.

6.1.1 The Significance of Inter-Domain Management

PREPARE was planned in 1991 and started its work in 1992, at a time when the area of cooperative end-to-end service management received less attention. However, with the increasing movement towards open telecommunications markets and the demand for global telecommunications services this situation changed considerably during the lifetime of the project. Awareness of the issues that need to be approached and solved in order to bring about the global information society has increased.

Over the last years, we have seen a rapid growth in networks and connectivity options, accompanied by deregulation, increased network usage, and new service offerings. These developments and new technologies are turning the former single service networks and privately owned networks into a globally shared integrated broadband communications infrastructure providing a multiplicity of services.

In terms of ownership and administration, this new communications infrastructure inherently consists of multiple management domains. However, the availability, reliability, interconnectivity, and service interworking of the communications infrastructure must be achieved and, in many cases, guaranteed. This has created the necessity for cooperation between network operators, service providers, and customers, and has also increased the demand for standardised management interfaces to support the end-to-end management of global services.

On one hand, the service providers' management systems need to cooperate in order to implement global telecommunications services and to leverage their infrastructure investments by providing a full set of advanced teleservices to their customers on a one-stop shopping basis. Since no service provider is able to equally well cover all geographic regions and telecommunications service offerings, not only the establishment of alliances and partnerships but also the deployment of open services and standardised inter-domain service management interfaces is of vital importance.

On the other hand, the customer organisations are asking for integrated private and public network solutions, such as end-to-end VPNs and VPN management, to support their corporate communications requirements. This also applies to individual end users who want to buy an increasing number of services from private and public providers in order to make efficient use of the information and capabilities supported by the broadband infrastructure. Such integrated private and public network solutions require public networks to be capable of providing the level of flexibility and management control that is normally associated with private networks and personal computers. As a consequence, a new set of requirements from customers and end users necessitating inter-domain management cooperation have to be supported by service providers.

In the future information society, inter-domain management is used to establish the basis for selecting between service providers, for on-line subscriptions to services, for customer or end user specific service configurations, and for on-line monitoring and control by the service users. For the service providers, inter-domain management enables global cooperation and service interworking as well as service creation and execution based on underlying supporting service offerings from other providers, leading to a vast set of new basic and value-added services for the information society.

6.1.2   Inter-Domain Management and TMN

As a consequence of the ONP-based approach, PREPARE worked on the assumption that in an open telecommunications service market there will be multiple cooperating service providers utilising a common, shared network infrastructure. Open standardised interfaces between the service providers' and infrastructure owners' respective management systems are a prerequisite to the effective, automated creation, provision, and operation of advanced services in such a market.

Today, there is not much choice about how to realise open management interfaces. The customers of advanced telecommunications management systems are the traditional telecommunications administrations (public network operators). There is a long tradition in the community of PuNOs of jointly defining, via the ITU, international standards which are needed to achieve network interoperability.

Realising that management networks (systems) also need to interoperate has led the PuNOs to define the TMN recommendations to ensure management system interoperability as well as interoperability between the management systems and the telecommunications systems. In this section we summarise the most important assumptions, interpretations, and conclusions about the TMN recommendations made by PREPARE.

6.1.2.1   PREPARE's Assumptions and Interpretations

PREPARE assumed that TMN principles are valid not only for traditional administrations (i.e., PuNOs) but also for other organisations such as the customers and users of the telecommunications services, as well as for private network operators (i.e., operators without supply obligations). Thus, each individual organisation in the organisational situations examined by PREPARE had its own TMN.

The M.3010 recommendation does not convey a very precise view about when something is a TMN [M.3010]. The interpretation by PREPARE is, in a simplified view, that a management system with an interface to other management systems that conforms to the TMN X interface definition in M.3010 is a TMN[1]. Until ITU defines conformance requirements for TMN this assumption is valid and does not contradict the TMN principles.

This view is also consistent with TMN's definition of the X interface, which is broad enough not to restrict the application of X interfaces to "pure" TMN-TMN interworking. It additionally allows the interworking of TMNs with any type of management system[2] as long as the interface principles defined in M.3010 are followed. Therefore, X interfaces would also apply to CPN management systems as well as to other, smaller network management systems.

6.1.2.2   Application of TMN Architecture Principles for Inter-Domain Management Interoperability

With respect to TMN's distinction between a functional and a physical architecture, this was found less useful, at least from an inter-domain management, X interface related perspective. The X interface is defined to exist between two physical building blocks containing OSFs or OSF-like function blocks [M.3010]. In practice, however, when developing the PREPARE communications management demonstrators, X interfaces were thought of as delineating TMNs as such, and what lay behind the interface was not of importance for the specification of the interface. Thus, there was not much to gain from distinguishing between reference points and interfaces. This was due to simplification attempts which resulted in defining as few interfaces as possible. In order to reduce complexity, only one X interface was defined between any two TMNs which had to interoperate. For intra-domain management concerns, other considerations may result in the application of both the functional and the physical architecture principles.

With respect to TMN's logical layered architecture (LLA) principle, it was found very useful to adopt a standard principle for the separation of management scope and responsibility in the internal structuring of a TMN. Based on a hierarchical control-feedback view of management, which is a natural approach in the management of telecommunications networks, the LLA can be used to define increasing levels of abstraction above the actual telecommunications resources. By specifying a hierarchy of OSFs in a TMN and by defining management layer-specific information models between these OSFs, the administrative authority over the TMN is able to control the external access by other service providers and service customers by implementing such external access at an abstract level, thereby separating the external requests from the actual resources and allowing the authority to enforce its own policies.

---

[1]   With respect to protocols it was assumed that those defined for Q3 interfaces apply equally well to X interfaces. This is a common and justifiable assumption [NMF-TR115].

[2]   "The X interface ... will be used to interconnect two TMNs or to interconnect a TMN with other network or systems which accommodates a TMN-like interface." [M.3010/95].

"The x reference points are located between the OSF function blocks in different TMNs. Entities located beyond the x reference point may be part of an actual TMN (OSF) or part of a non-TMN environment (OSF-like). This classification is not visible at the x reference point." [M.3010/95].

At the same time, this allows simple-to-use, high-level abstractions of the actual resources to be defined to suit the needs of the users of the management services provided by the TMN. Here it should be noted that PREPARE regarded inter-domain management as primarily a concern for the service layer of management. The organisations which cooperated via inter-domain management were all using different services and so the management information they exchanged was service management information.

From the start of PREPARE it was found that the TMN principles and OSI Systems Management did not provide support for some inter-domain management-specific activities and technical needs. The solution was an X.500 directory-based concept which was implemented in the inter-domain management demonstrators (see Appendix C). PREPARE found that the requirements identified were valid TMN architectural requirements and that the solution implemented was adequate to satisfy the requirements within the framework of the TMN principles. The natural consequence was to contribute these to ITU which accepted the proposal for an enhancement of the TMN functional architecture. This has implications for a corresponding enhancement of the TMN information architecture (see section A.4).

The experience of PREPARE shows that the TMN prescriptions for the specification of TMN interfaces are sufficient for specifying implementable, interoperable inter-domain management interfaces (cf. chapter 3). The TMN interface specification methodology covers all the actual specification needs [M.3020/94]. As a documentation style the NMF Ensemble format is appropriate for documenting such interface specifications [NMF-025]. From a methodology point of view, however, it does not cover sufficiently the requirements definition stage for inter-domain management purposes. PREPARE developed its own approach based on RM-ODP enterprise language concepts.

6.1.2.3   TMN Incompleteness

The interpretation of TMN that interfaces must be defined in standards/recommendations in order to qualify as TMN interfaces implies that today's TMN recommendations are still incomplete from the perspective of actual X interface specifications. In particular, TMN service layer management information models, which are becoming increasingly important, are not (yet) defined as part of any ITU recommendation.

The use of management technologies and standards other than OSI Systems Management, which underlies TMN, is a reality today and there is no sign that in the future the number of technologies and standards domains will decrease (cf. chapter 5). For example, SNMP is, for various reasons, the preferred approach in smaller, private and corporate networks such as LANs. It would appear to be a problematic constraint on the applicability of TMN that it prescribes, at an architectural level, the protocols and information specification techniques to a subset of available and utilised options.

The adoption by ITU of PREPARE's proposed enhancements of the TMN functional architecture is, however, taken as evidence that the TMN recommendations are open not only in offering internationally agreed, publicly available specifications, but also that both the process of defining the standards as well as the standards themselves are open. Therefore there seems to be nothing which prevents TMN from evolving in those directions in the future.

The conclusion with respect to TMN is that even though it does not, in its present state, address all aspects of inter-domain management and all emerging new technologies and trends, it is still an adequate basis for a management architecture for an open service market. In particular, TMN enables integrated service management including pre-service management activities and cooperation between provider and customer management systems.

6.1.3   Implementation Experiences

As part of the realisation of the two demonstrators, PREPARE adopted the TMN framework for the definition and implementation of the inter-domain management systems and, in particular, for the inter-domain management interfaces, which are TMN X interfaces. The practical experiences gained during implementation of the inter-domain management demonstrators are described in the following.

6.1.3.1   TMN Development and the Runtime Environment

First of all, the implementation of TMN systems requires powerful platform support. In order to facilitate efficient agent and management application development the design of GDMO and ASN.1 specifications of TMN interfaces needs to be supported by appropriate editors, syntax checkers, and class browsers. The final specifications are processed by tools generating implementation source files to be completed manually by the implementers. The platform runtime environment consists of libraries and application programming interfaces (APIs) that fit with the generated code so that the application developers do not have to be concerned with handling, for example, CMIP request and event queues, ASN.1 encoding/decoding, association establishment and release, etc.

The APIs needed on the management platforms should be "high-level APIs" that allow management application development in an object-oriented environment (e.g., C++). In particular, they should cover various aspects of management applications, such as transparent communication between manager and agent applications, set-up and release of associations, access to real resources, object-to-object communication, proxy agent implementation, etc. In the case of inter-domain management, communication support for global naming and addressing, distribution transparency, authentication, and access control are also vital requirements on management platforms. Meanwhile, there are several standards groups working on the definition of such APIs, such as the CMIS API working groups within the NMF and X/Open [NMF-CMIS++].

Not all of the platform support briefly sketched above was available to all PREPARE partners during the project. However, the important requirements for TMN platform support were experienced and communicated to product groups inside and outside the project. This led to new deliveries and improvements to TMN platform technology during the project both within and outside the PREPARE partner organisations.

6.1.3.2   Platform Interoperability

Beside the management platform support in the form of development tools and APIs, one of the greatest problems in the implementation was to achieve interoperability between the various management platforms that were used in the project in the different domains. As the demonstrator implementations were realised on several interworking platforms supplied by different platform providers, some consideration had to be given to platform interoperability tests before the full integration could be planned. A specific test scenario was chosen based on the NMF's test managed object. This scenario was then run remotely between the respective platforms. This exercise provided the means of solving the basic interoperability problems between the platforms, but in a multi-platform environment even such simple interoperability tests were time-consuming.

Among the most difficult interoperability problems were the set-up and release of associations and the handling of notifications. There were various reasons for the problems. For example, the different platforms used different ways to address remote management applications. This meant that they used different addressing concepts and different mechanisms to forward incoming management requests and replies to several management applications in one system. Some platforms use Application Entity Titles (AETs) for the manager and agent applications, some use AETs just for agent applications, and others do not use titles to establish new associations.

If one platform did not use an AET for addressing but the other platform did, it was impossible to specify the address of an existing management application in an Event Forwarding Discriminator (EFD) in the expected way. The agent received only communication address information, but not information on how to address the management applications. The only solution found for the first demonstrator was to avoid configurations where the problem occurred. For the second demonstrator, updated versions of all platforms used in PREPARE could be adopted which provide modified addressing of management applications.

All these initial problems and the effort required to achieve proper interworking and cooperation between the TMN management systems from different domains showed that TMN is a complex technology. However, the demonstrators also proved that cooperative inter-domain management is possible with TMN and that complex actions, such as end-to-end reservation and set-up of virtual paths, alarm propagation between service providers and to customers, on-line configuration, and billing, etc., can be implemented.

6.2   Areas Requiring Further Work

Despite the increased importance of inter-domain management for IBC in an open service market there are still many open issues. As a pilot project for advanced end-to-end resource management PREPARE explored and extended available technical frameworks and applied them in practice. Some of the main areas that were found to require substantial further research, development and, in particular, standardisation efforts are addressed in this section.

6.2.1   Inter-Domain Management Standards

The work of PREPARE provides evidence that the specification of inter-domain management interfaces and the realisation of inter-domain management systems based on these specifications is feasible. For inter-domain management to provide the required basis for the cooperation and interworking of different players in an open service market, however, many more open standards and broad implementations of these standards are needed.

One of the areas needing a universally applicable solution is global naming and addressing of services, systems, and resources. PREPARE used the X.500 Directory integrated with OSI Systems Management standards for that purpose. The requirements to support global naming and addressing were also contributed to the ITU and are now explicitly present in the TMN recommendations (M.3010) as a Directory Service Function (DSF) and Directory Access Function (DAF) (cf. section A.4). The viability of this solution depends on the acceptance of X.500 and its worldwide deployment and usage. Other common means to provide name services in distributed systems today, such as the Domain Name System used in TCP/IP networks or the envisaged solution for the global naming of CORBA objects, need to be integrated as alternatives to provide a high degree of flexibility for end-to-end cooperation.

As with global naming and addressing, the choice of protocols and services for transporting management information should not be restricted to CMIS/CMIP and FTAM as in today's TMN recommendations for Q3 and X interfaces. Worldwide interaction and cooperation between systems that are separately administered is shown to work every day on the World Wide Web using, amongst other types of service, hypertext and simple file transfer.

The management services and information available at inter-domain management interfaces need to be standardised, however. This will require ongoing work to fill the gap for service layer management information models, which is one of the major hurdles to facilitating inter-domain management today. On one hand, generic service management operations and object models are needed to enable the efficient realisation of service portfolios and interworking between providers. On the other hand, refinements of the generic model and instantiation of management information models for particular service offerings are required to exploit the strengths of individual end-to-end service offerings. Currently, one of the most active groups working on these issues is NMF's SMART (cf. section 5.4).

Probably the most important difference between intra-domain and inter-domain management activities is the need for security at inter-domain management interfaces. Due to the fact that inter-domain management interfaces are inherently for use by external parties, usage permission must be explicitly granted and the identity of the management service users must be verified by strong authentication procedures. External access to pieces of the management information visible at inter-domain management interfaces must be controlled by the service provider's agent function. This is another important area of standardisation.

Meanwhile, many of the open issues mentioned above have been recognised and are being worked on in, for example, ITU-T's TMN Expert Group, Question 5/4. First drafts of a recommendation on the TMN X interface (M.Xreq) are being produced. The goal of future inter-domain management standards must be to make multi-domain management efficient enough to be comparable to intra-domain management in a single management administrative domain.

6.2.2   Open Management Platform Support

Provider-provider and provider-customer cooperation in an open service market cannot be achieved with platforms that use proprietary protocols and management information data formats. This has been widely recognised during recent years but needs to be stressed again. The various vendor platforms need to open up and support the standards and remain open and extensible. This is particularly valid for APIs since they are the key to efficient development and operation of multi-domain management systems. In addition, common APIs improve the interoperability and portability of management applications. Therefore, future management platforms should be measured and evaluated with particular attention paid to the existence of and degree of support for common high-level APIs[1].

PREPARE's implementation experiences also point directly to the need for development tool support as described in section 6.1.3.1. Whereas development will have to take place on powerful workstations, running the actual inter-domain management applications is a different matter. Here, for example, scalability of TMN systems according to a client/server organisation of processes and roles is very important. Large operations systems servers are needed to run a PuNO's integrated services broadband network, but individual partner providers and customers may also access the TMN via small management application clients running on smaller workstations and personal computers. Such requirements demand the scalable distributed realisation of multi-domain management systems on open platforms.

The activities in the TINA consortium (TINA DPE) and NMF's SPIRIT, as well as OMG's CORBA are important initiatives in this direction (cf. chapter 5).

6.2.3   Integrated Service and Systems Management

More and more of today's service offerings on an integrated broadband communications infrastructure are realised in software. Multimedia teleservices as used in PREPARE, such as mailing and conferencing, application sharing, etc., represent distributed service software. The client/server parts of these distributed applications run on powerful workstations attached to switches somewhere in the broadband network. In an open service market new service offerings of this kind have to be created and deployed rapidly and there is less and less time for the service providers to recover the development costs. To provide the services themselves, complex real-time server systems have to be set up, such as for video-on-demand, tele-cooperation, etc.

The trends described above, the ongoing integration of voice and data applications, computer-telephony integration, real-time and broadband applications, etc., all create a strong demand for the full integration of service management and systems management. Despite these trends, the techniques used today for service management and systems management still differ. The influence of new approaches to service management and control in provider networks, i.e., the TINA initiative, and emerging distributed object-oriented frameworks, i.e., OMG/CORBA, are also promising improvements to this situation.

In order to facilitate the integration of service and systems management incorporating today's installed base of management platforms and applications, management gateways between different protocols, frameworks and information models ("standards domains", cf. section 5.7), such as the CMIP/SNMP gateway used in PREPARE, have to be implemented. From an architectural point of view, many of the desktop end user applications will be moved into the network and provided as networked services

---

[1]   Compare, for example the work being undertaken by the Network Management Forum and X/Open to provide an overall framework for a set of APIs for standards based network management [NMF-TMN].

and applications. Accordingly, the management of these applications needs to be adapted to accommodate time and location, as well as administration and policy differences.

It is the task of integrated service and systems management in an open service market to provide distribution transparency as much as possible and to ensure the provision of end user services over a seamless network.

6.3   Outlook

There is no indication that in any foreseeable future we shall see a dramatic reduction in the number of telecommunications infrastructure owners and service providers. On the contrary, the full liberalisation of the European telecommunications service market, which will come into effect by 1st January 1998 at the latest, will lead to many new providers, e.g., railway and electricity supply companies. Therefore, the concept of inter-domain management will be of high relevance in the years to come.

6.3.1   TMN and Future Technologies

As explained in this book, TMN was used as the architectural base standard for the inter-domain management development work in PREPARE [M.3010]. While the TMN principles have not evolved much during the lifetime of PREPARE, much activity has taken place in the area of distributed computing.

TMN is usually thought of as a technical specification. Evidently, most specifications are associated with the detailed specification of protocols, management services, and information models. However, TMN is more than that and the key role of TMN, of which the actual interface specifications are just one visible result, is more important than as a technology for telecommunications management systems. TMN must be recognised as a political statement by telecommunications management system customers as to how the systems they need have to interoperate (the property of technical openness). Hence the term telecommunications management *network*.

It is important here to note that telecommunications management is not a purely technical or functional problem. In a commercial marketplace (i.e., an open service market) telecommunications management is about responsibility, guarantees, commitment, connectivity, control, and autonomy. The TMN is both a conceptual network and instrumental in the implementation of the other concerns. Having realised that, we have to consider what the TMN recommendations aim to define and, equally important, what they do not define. It is in the light of such an assessment that emerging architectures and technologies should be judged.

Besides TMN, the most serious influence is expected from the TINA Consortium. As a simplified characterisation of the two initiatives, TMN addresses the problem of interoperability of physical management systems by defining open interfaces (based on OSI standards), whereas TINA addresses the design and deployment of telecommunications service software, including management software.

The reasons for this new type of software-specification standards work, as represented by TINA-C, is the realisation that telecommunications services and management services are to an increasing extent software-based (as envisaged with the IN standardisation effort). Moreover, it can be seen as a realisation of customer wishes for a higher degree of vendor independence and a maximum degree of openness and thus for freedom of choice with respect to physical systems, independent software vendors, manufacturers, etc.

With respect to TINA, future telecommunications services have the following core characteristics:

(1)   They need to be created rapidly, sometimes under customer control.

(2)   They need to be reusable and "combinable" in order to enable (1).

(3)   They need to be modular to support (2).

A key enabler is distribution transparency. For that purpose TINA-C defines a common, technology-independent platform specification (DPE) [TINA-005]. This supports all well-known requirements and characteristics of (open) distributed processing, as defined in the RM-ODP. As the major PuNOs (as well as some computer/software houses) participate in TINA, it is foreseeable that their specifications will, at some point in the future, become part of the tenders and procurement specifications of these PuNOs.

TMN and TINA are therefore both to be considered as "universes of discourse" defining the world view existing between customers and suppliers of telecommunications management systems and software, and also as a statement of longer term goals for openness (vendor-independence, interoperability). At an abstract level, the main difference between TMN and TINA can be characterised by their respective views on what constitutes the "unit of procurement". In TMN, these are physical systems (physical building blocks in TMN terminology), whereas in TINA they are individual software components which can be seen as a finer granularity level compared to TMN.

The emergence of commercial CORBA conformant products is also expected to have implications for inter-domain management system development. TINA assumes a common universal service infrastructure, the DPE, for which CORBA is seen as an intermediate implementation option. The NMF has outlined an integration architecture in which the CORBA IDL is recommended as the interface definition notation for management solutions in a multi-technology environment (where TMN and SNMP are seen as technologies) [TR114]. CORBA is also part of NMF's platform procurement specification SPIRIT [SPIRIT].

It would be naïve to be believe that these new and promising technologies and approaches will be available in the near term to be employed for inter-domain management. It may take several years to complete the specifications and implementations required before practical application is possible, similar to the development of the TMN framework. Their suitability and maturity will have to be proved in the years to come. In the meantime, the number of TMN installations and inter-domain management interfaces will increase, as will the experiences gained using TMN.

6.3.2   PROSPECT

Recognising the importance of pre-competitive research and the necessity of field trials to gain practical experiences with provider-provider and provider-customer interactions in an open service market, the PREPARE partners have started a follow-on project called PROSPECT under the EU ACTS framework. The goal is to run several trials with end users and a variety of service providers, amongst them new partners, with each trial lasting several months. Various multimedia teleservices are offered via a pan-European broadband network infrastructure. Inter-domain management facilitates the flexible use of the services by the end users and their reuse by service providers in order to build new services from existing building blocks. For example, a tele-educational service offering which is part of the trial relies on multimedia mailing and conferencing, as well as on application sharing and hypertext information systems.

Within PREPARE, TMN was used to demonstrate end-to-end management including private and public networks and covering the network element management, network management, and service management layers. The latter, service management, was mainly seen from the perspective of broadband bearer service examples, such as ATM and also VPN. For the second demonstration, multimedia teleservices were included, but this just provided an insight into future necessary inter-domain management functions in an open service market.

Within PROSPECT, the management of enhanced teleservices and the service management infrastructure required to support open interfaces and cooperation between different kinds of service provider mutually using and building on each other's services is the central topic. Key questions to the success of PROSPECT are how to combine and integrate technologies such as TMN for the network and bearer service infrastructure and, for example, SNMP or CORBA for systems and enhanced service management. Through field trials with providers and end users we have to learn more about the potential, required design, and acceptance of multimedia teleservices and their management on the information highways.

A Telecommunications Management Network (TMN) Recommendations

A.1 Motivation for Standardising TMN

The ITU concept of a TMN was introduced in the mid-80s in CCITT recommendation M.30 *Principles for a TMN*. In the 1992 edition M.30 was renumbered and is now known as ITU-T recommendation M.3010 [M.3010].

TMN was originally envisaged as a network of interoperable management systems which could provide integrated management of telecommunications networks. This network of interoperable management systems comprises TMN. It was realised that standards were required in order to ensure interoperability between the management systems. Since the publication of M.30 in 1988 several standards bodies, in particular ITU and ETSI, have been working on aspects of the TMN framework.

A.2 Overview of TMN Recommendations

The following table gives an overview of the TMN recommendations currently available from ITU.

ITU-T Rec.	Title
M.3000	OVERVIEW OF TMN RECOMMENDATIONS
M.3010	PRINCIPLES FOR A TMN (see below)
M.3020	TMN INTERFACE SPECIFICATION METHODOLOGY (see section 3.7.4.1)
M.3100	GENERIC NETWORK INFORMATION MODEL
M.3101	CONFORMANCE STATEMENT PROFORMAS FOR RECOMMENDATION M.3100 "GENERIC NETWORK INFORMATION MODEL"
M.3180	CATALOGUE OF TMN MANAGEMENT INFORMATION
M.3200	TMN MANAGEMENT SERVICES: OVERVIEW
M.3300	TMN MANAGEMENT CAPABILITIES PRESENTED AT F INTERFACE
M.3400	TMN MANAGEMENT FUNCTIONS
M.3640	MANAGEMENT OF THE D-CHANNEL DATA LINK AND NETWORK LAYER
M.3641	MANAGEMENT INFORMATION MODEL FOR THE MANAGEMENT OF THE DATA LINK AND NETWORK LAYER OF THE ISDN D-CHANNEL
G.773	PROTOCOL SUITE FOR Q-INTERFACE FOR MANAGEMENT FOR TRANSMISSION SYSTEMS
G.774	SDH MANAGEMENT INFORMATION MODEL FOR THE NETWORK ELEMENT VIEW

Q.811	LOWER LAYER PROTOCOL PROFILES FOR THE Q3-INTERFACE
Q.812	UPPER LAYER PROTOCOL PROFILES FOR THE Q3-INTERFACE
Q.821	STAGE 2 AND STAGE 3 DESCRIPTION FOR THE Q3-INTERFACE-ALARM SURVEILLANCE
Q.822	STAGE 1, STAGE 2 AND STAGE 3 DESCRIPTIONS FOR THE Q3-INTERFACE-PERFORMANCE MANAGEMENT

Table A.1: ITU TMN Recommendations

Each of these recommendations is briefly described in *Overview of TMN recommendations* [M.3000]. In addition, the European Telecommunications Standardisation Institute (ETSI) is producing an ETSI Technical Report entitled *Network Aspects - Telecommunications Management Network (TMN) - TMN Standardisation Overview* [ETR230] which provides a comprehensive overview of TMN recommendations and standards from ITU and ETSI.

A.3   M.3010: Principles of a TMN

The TMN architectural framework considers four aspects of a management network (i.e., a TMN) [M.3010]. The first three aspects are as follows: the functionality it contains specified in terms of a *functional architecture*; the physical nodes it contains specified in terms of a *physical architecture*; and the management information which is communicated between pairs of function blocks or pairs of physical blocks (nodes) of the TMN, which is called the *information architecture*. The functional and physical architectures specify how TMNs can be constructed from TMN building blocks with interfaces between them. These concepts (TMN building blocks and interfaces) have different names in the two architectures (as described below). The information architecture describes how management information is to be modelled and specified. The fourth aspect is a hierarchical structuring of management responsibility known as the *Logical Layered Architecture* (LLA). The LLA is a generic principle which in M.3010 is exemplified in combination with the functional architecture, based on the concepts of business, service, network, and element management layers[1].

A.3.1   TMN Functional Architecture

The functional architecture describes the appropriate distribution of functionality within the TMN to allow for the creation of function blocks from which a TMN of any complexity can be implemented. The elements of the functional architecture are *function blocks* and *reference points*. Function blocks are conceptual entities that can be implemented in a variety of physical configurations. Reference points represent the exchange of information between pairs of function blocks. Functions generally needed in the context of communicating management systems and applications have been defined as a set of "generic" function blocks.

---

[1]   In the 1992 edition of M.3010 the LLA example was described by a combination of the TMN functional architecture and the TMN information architecture [M.3010]. In later versions only the TMN functional architecture was used as an example [M.3010/95].

M.3010 defines the types of function block and reference point, of which the following are particularly relevant to this book:

Operations System Function (OSF) block. The OSF processes information related to telecommunications management for the purpose of monitoring/coordinating and/or controlling telecommunications functions, including management functions (i.e., the TMN itself).

Network Element Function (NEF) block. The NEF is a function block which communicates with the TMN for the purpose of being monitored and/or controlled. The NEF provides the telecommunications and support functions which are required by the telecommunications network being managed. The NEF includes the telecommunications functions which are the subject of management. These functions are not part of the TMN but are represented to the TMN by the NEF. The part of the NEF that provides this representation in support of the TMN is part of the TMN itself, whilst the telecommunications functions themselves are outside the TMN.

Workstation Function (WSF) block. The WSF provides the means to interpret TMN information for the human user, and vice versa. The responsibility of the WSF is to translate between a TMN reference point and a non-TMN reference point and hence this latter activity is outside the TMN boundary.

Q Adaptor Function (QAF) block. The QAF block is used to connect as part of the TMN those non-TMN entities which are NEF-like and OSF-like. The responsibility of the QAF is to translate between a TMN reference point and a non-TMN (e.g., proprietary) reference point and hence this latter activity is outside the TMN boundary.

The *q reference points* are located between the function blocks NEF and OSF; NEF and MF[1]; MF and MF; QAF and MF; MF and OSF; QAF and OSF; and OSF and OSF, either directly or via the data communication function (DCF). The following are within the class of q reference points:

q_x The q_x reference points are between NEF and MF, QAF and MF, and between MF and MF.

q_3 The q_3 reference points are between NEF and OSF, QAF and OSF, MF and OSF, and OSF and OSF.

The q_3 and q_x reference points may be distinguished by the knowledge required to communicate between the function blocks they connect. The distinction is for further study.

The *x reference points* are located between the OSF function blocks in different TMNs. Entities located beyond the x reference point may be part of an actual TMN (OSF) or part of a non-TMN environment (OSF-like). This classification is not visible at the x reference point.

Two OSFs interoperate via a q reference point when both OSFs are located in the same TMN and via an x reference point when the two OSFs are located in different TMNs, as depicted in Figure A.1 (a).

---

[1]　MF is the TMN *Mediation Function* block which is able to mediate between two standard TMN interfaces.

The *f reference points* are located between the WSF and the OSF blocks and/or the WSF and the MF blocks.

Each function block is itself composed of *functional components*. Standardisation of functional components is, however, outside the scope of TMN recommendations. The following are examples of functional components (see also section A.4):

Security Function (SF). The security functional component provides the security service that is necessary for function blocks to satisfy the security policy and/or user requirements.

Message Communication Function (MCF). The MCF is associated with all functional blocks having a physical interface. It is used for, and limited to, exchanging management information contained in messages with its peers. The MCF is composed of a protocol stack that allows function blocks to be connected to Data Communication Functions.

User Interface Support Function (UISF). The UISF translates the information held in the TMN information model into a displayable format for the human-machine interface, and translates user input into the TMN information model.

Management Application Function (MAF). An MAF represents part of the functionality of one or more TMN management services. In order to carry out TMN management services, interactions take place between MAFs in different function blocks with the help of other functional components. Each interaction, known as a TMN management function, involves one or more pairs of cooperating MAFs. An MAF includes a logical representation of the management information exchanged with other MAFs.

A.3.2   TMN Physical Architecture

The elements of the physical architecture are building blocks and interfaces. Building blocks are the different types of physical node in the TMN, whereas interfaces define the information exchange between them. The physical architecture describes realisable interfaces and generic examples of physical components that make up a TMN. *Generic* physical building blocks and interfaces are defined in M.3010. These building blocks generally reflect a one-to-one mapping (or implementation) of one function block to one physical building block. For example, an operations system (OS) physical building block contains one (or more) OSFs. This does not prevent an actual TMN implementation from containing multiple different function blocks in one or more physical blocks, nor does it prevent the distribution of one function block over several physical nodes. Interfaces are the implementation of reference points from the functional architecture. In particular, a Q interface implements a (set of) q reference point(s), whereas an X interface implements a (set of) x reference point(s), as depicted in Figure A.1 (b).

A.3.3   TMN Information Architecture

The information architecture describes an object-oriented approach for transaction-oriented information exchange within a TMN. This comprises a management information modelling aspect and a management information exchange aspect, both of which are adopted from the OSI standards (see Appendix B).

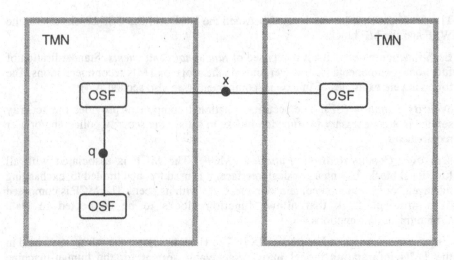

(a) Illustration of aspects of TMN Functional Architecture

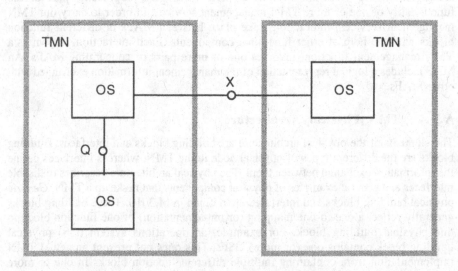

(b) Illustration of aspects of TMN Physical Architecture

Figure A.1: Illustration of aspects of the TMN architecture

The TMN in its current state is based on the management information communications principles defined in the context of the OSI Systems Management standards which are being standardised jointly by ITU and ISO/IEC. ITU recommendations *Management Framework for OSI* [X.700] and *Systems Management Overview* [X.701] define an architecture for the exchange of management information between (peer) open systems (i.e., systems with OSI conformant communications capabilities). Appendix B gives an introduction to OSI Systems Management; here we just note that its basic constituents, which have been adopted in the TMN standards, are the communications protocols and the object-oriented representation of management information. The latter constitutes the basis for the TMN information architecture. OSI Systems Management introduces the concept of Managed Object (MO) which is defined as "the management view of a resource that is subject to management". Moreover, OSI Systems Management identifies systems management as being a distributed application in which communicating application processes in their interactions assume the role of either manager (client) or agent (server). These roles are not static but may change from one service invocation to the next.

A.3.4 The Logical Layered Architecture

The fourth TMN architectural principle is the *Logical Layered Architecture* (LLA). The LLA is a concept for structuring management functionality into a grouping called "logical layers" and describes the relationship between layers. A logical layer reflects particular aspects of management. Even though the LLA principle is very generic, it is most often identified with the example grouping of telecommunications management into the four hierarchical layers of business management, service management, network management, and network element management. These layers view management at different abstraction levels such that *business management* is concerned with overall enterprise management, *service management* is concerned with managing customer services, *network management* is concerned with managing individual networks, and *network element management* is concerned with managing individual components of a network. This LLA principle is often combined with the functional architecture to form principle architectures of "real TMNs". For instance, in Figure A.1 (a), the two function blocks in the leftmost TMN could be a "service layer OSF" at the top, and a "network layer OSF" at the bottom.

A.4   TMN Model of the X.500 Directory

The 1996 version of ITU recommendation M.3010 includes revisions in the areas addressed by PREPARE, namely inter-domain management across TMN X interfaces [M.3010/95]. An appropriate extension to the TMN architectural framework has been introduced based on a generalisation of PREPARE's experience in architectural and information modelling (see chapter 3). This section briefly outlines the extension.

TMN can benefit from the general features of the X.500 directory, such as general purpose directory information, as well as from specialisations for the purpose of TMN interoperability. The latter requires the specification of TMN-specific directory information. Since some parts of the directory system are outside any TMN whereas other parts will be within TMNs, there is a need to model the distribution of the directory to the extent that it affects the TMN interfaces. This was done by adding directory capabilities to existing TMN interfaces and by adding new functional components to the TMN architecture.

A differentiation is needed between access function and provision (system and information repository) function. This is in accordance with the X.500 recommendations since there is no requirement to know about the actual location of particular objects needed when applying the access function [X.500]. Users of the directory services may issue requests to any provision function and the directory system will locate and provide the relevant objects/information. This means that provision functions need to be interconnected within TMNs, across TMNs, and across TMNs and non-TMN systems. In accordance with the diverging requirements of the two capabilities of directory access and distributed directory provision the distinction between the two capabilities is required.

The directory was introduced into the TMN architecture by new functional components representing directory information and service access capabilities termed *Directory Access Function* (DAF), and directory system provision capabilities termed *Directory System Function* (DSF).

A.4.1   Directory Access Function

The DAF is associated with all TMN functional blocks which need to access the directory (mainly required for the OSF, but possibly also useful for other function blocks). It is used to obtain access to and/or maintain (read, list, search, add, modify, delete) TMN-related information represented in the Directory Information Base (DIB). An X.500 directory system is one candidate for implementing the required directory functionality. In X.500 the DAF is termed Directory User Agent (DUA). It provides access to the directory, or more specifically to a Directory System Agent (DSA), via the Directory Access Protocol (DAP).

A.4.2   Directory System Function

The Directory System Function component represents a distributed directory system that is available locally or globally. As an implementation option it could be built by one or more DSAs as described in the X.500 series of recommendations. Each DSF stores directory objects (DOs).

A.4.3   Access to the X.500 Directory

TMN function blocks may optionally use X.500 directory functional components to implement the required directory function. Figure A.2 illustrates how the interworking between the TMN and the directory environment is modelled.

A DAF may be integrated in each TMN function block to allow access and maintenance of TMN-related information in the DIB. The DSF functional components of the directory may be associated with DAF functional components within one TMN for local directory support (access via q reference point); they may be associated with functional blocks in remote TMNs (access via x reference point); or they may exist in non-TMN directories (access via an x reference point to an OSF-like block outside the TMN environment).

Figure A.2 shows an example illustrating the options for using and providing directory services and information in the TMN architecture. It shows two TMNs (A and B) and the general purpose directory. In TMN (A) there are three TMN function blocks, two containing DSF functional components (DSF_a and DSF_b), and one containing the DAF functional component. In TMN (B) there is shown one TMN function block containing the DSF functional component DSF_c. In the directory there is shown a non-TMN function block containing a DSF TMN functional component (DSF_d). Each of the DSFs holds a DIB fragment, i.e., a collection of directory objects.

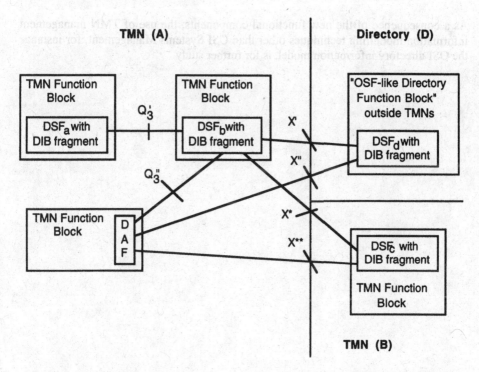

Figure A.2: Relationship between the TMN and directory components

For a user of the DAF functional component in TMN (A) there are three ways of accessing information held in the directory:

1. To issue requests to the TMN function block containing the DSF_b functional component over the interface Q_3''.

2. To issue requests to the general purpose directory holding the functional component DSF_d, over TMN interface X".

3. To issue requests to the TMN function block holding the functional component DSF_c in TMN (B) over interface X**.

In no case is there a requirement on the corresponding DSF functional component to store locally the actual directory object / information requested. It may be retrieved by the distributed operations of the directory, by inter-DSF communications, as follows:

1. Between DSF_b and DSF_a, both located in building blocks in TMN (A) and therefore communicating over Q_3'.

2. Between DSF_b and DSF_c, located in building blocks in different TMNs and therefore communicating over X*.

3. Between DSF_b located in a building block in TMN (A) and DSF_d located in the general purpose directory and therefore communicating over X'.

As a consequence of the new functional components, the use of TMN management information modelling techniques other than OSI Systems Management, for instance the OSI directory information model, is for further study.

B   OSI Systems Management

B.1 Introduction

Network management concepts first appeared in the OSI *Reference Model* [X.200]. This was later extended to include an OSI *Management Framework* [X.700] which identifies three categories of management as shown in Figure B.1:

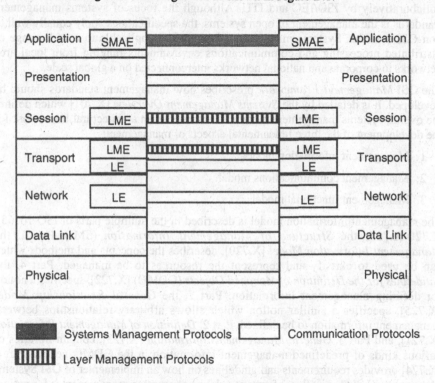

Figure B.1: OSI management structure

- *(N)-layer operations* are the control mechanisms (exceptions, aborts, etc.) already provided by ordinary (N)-protocols. This category allows the continued use of existing protocols to support specific layer activities. The (N)-layer operations manage a single instance of communication within one layer.

- *(N)-layer management* is concerned with the operation of a single layer. It can manage multiple instances of communication but can affect only resources belonging to the (N)-layer. (N)-layer management is accomplished by a *layer management entity* (LME) performing an (N)-management protocol which coexists with the normal (N)-protocol.

- *OSI Systems Management* is concerned with the monitoring and control of resources within remote systems. Not only resources belonging to all seven layers of the OSI Reference Model but also other resources (including non-OSI resources) may be managed in this fashion. The exchange of management information at this level is accomplished using a systems management protocol which is performed by a *Systems Management Application Entity* (SMAE).

Within ISO, standardisation in the fields of layer management and layer operations is regarded as a task for the respective layer working groups. Therefore, the management group (ISO/IEC JTC1/SC21/WG4) deals exclusively with OSI Systems Management. The ITU (formerly CCITT) decided to adopt ISO/IEC management standards and drafted the X.700 series of recommendations which are technically aligned with the ISO/IEC standards. New systems management standards are now developed collaboratively by ISO/IEC and ITU. Although the focus of systems management standards is the management of open systems, the specifications apply equally well to non-OSI systems. Systems management is therefore applicable to a wide range of distributed processing and communications environments ranging from local area networks to corporate and national networks interconnected on a global scale.

The OSI *Management Framework* prescribes how management standards should be developed. It is detailed by the *Systems Management Overview* [X.701] which defines the overall systems management model. This serves as an architectural framework for the development of the three fundamental aspects of management:

1. Management information model.

2. Management communications model.

3. Management functional model.

The management information model is described in the multiple parts of ISO 10165 / X.720-X.725, the *Structure of Management Information* (SMI). Part 1, the *Management Information Model* [X.720], describes the concepts and methods which can be used to classify and represent the resources to be managed. Part 4, the *Guidelines for the Definition of Managed Objects* (GDMO) [X.722], specifies a notion for defining management information. Part 7, the *General Relationship Model* [X.725], specifies a similar notion which allows arbitrary relationships between management information to be defined. Part 2, *Definition of Management Information* [X.721], and Part 5, *Generic Management Information* [X.723], contain libraries of various kinds of predefined management information in the GDMO notion. Part 6 [X.724] provides requirements and guidelines on how an implementer of OSI Systems Management has to specify conformance claims concerning management information. The information model is described in more detail in section B.2.

The management communications model includes the definition of the *Common Management Information Service* (CMIS) [X.710], the *Common Management Information Protocol* (CMIP) [X.711] as well as the associated protocol implementation conformance statements [X.712]. The communications model is described in more detail in section B.3.

The management functional model is described in the multiple parts of ISO 10164 / X.730ff (see Table B.1). A part may specify a set of systems management services and their mapping onto the underlying management communication services, generic definitions of management information, which are collected in Part 2 of SMI [X.721], and the definition of systems management functional units. More details about the functional model are given in section B.4.

B.2   Management Information Model

The OSI Systems Management information model is described in the multiple parts of ISO 10165 / X.720-X.725, the *Structure of Management Information* (SMI). This section focuses on Part 1, the *Management Information Model* [X.720] which includes the basic concepts.

B.2.1   The Managed Object Concept

The design principle of SMI is based on an object-oriented modelling approach. OSI management is carried out by accessing and manipulating *managed objects* (MOs). MOs are abstractions of data processing and data communications resources. An MO is an instance of a *managed object class* which is defined as a collection of packages, each of which is defined to be a collection of *attributes*, *actions*, *notifications*, and related behaviour (cf. Figure B.2). The distinction between the MO and the resource that it represents for management purposes may be described by stating that attributes, operations, and notifications are visible to management at the *managed object boundary* whereas the internal functioning of the resource that is represented by the MO is not otherwise visible to management.

Packages are either mandatory or conditional depending upon some explicitly stated condition that is evaluated when the MO containing the package is created. The value of an attribute can determine or reflect the behaviour of an MO. Actions are operations on an MO, the semantics of which are specified as part of the MO class definition. Notifications are emitted by an MO and contain information relating to an event that has occurred within the MO.

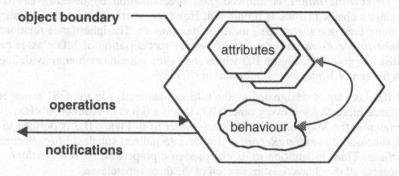

Figure B.2: Managed object concept

MOs are subject to management operations. Two types of management operations can be performed on MOs:

- *Object-oriented operations* manipulate an MO as a whole. This includes the creation and deletion of MOs. In addition, an *action* operation is available which is used to request the MO to perform the specified action and to indicate the result of that action. The action and optional associated information are part of the MO class definition.

- *Attribute-oriented operations* are sent to MOs to read or modify the values of attributes. According to their syntax, the attributes can be divided into *simple attributes*, *set-valued attributes*, and *attribute groups*. A simple attribute possesses a single value, while the value of a set-valued attribute is composed of an unordered, possibly empty, set of members of a given type. The size of the set is variable. An attribute group refers to a collection of attributes within an MO. An operation upon an attribute group is executed upon each attribute that is referenced by the group. i.e., all attributes that are to be operated upon by the operation request are considered to be available to the MO as part of a single request. The operations allowed on a given attribute and, optionally its required or permitted value set, are specified as part of the class definition of the encapsulating MO.

A *Management Information Base* is the conceptual repository of all MOs located within an open system.

B.2.2   Inheritance and Allomorphism

One MO class (the *subclass*) can be specialised from another MO class (the *superclass*) by defining it as an extension of the superclass. Such an extension is made by:

- Using the *inheritance* mechanism by which attributes, notifications, operations, and behaviour (subsumed under the term *characteristics*) are acquired by the subclass from the superclass.

- By extensions to the capabilities of the inherited characteristics.

- By the addition of new characteristics, such as new attributes, new management operations allowed on inherited attributes, etc.

Only *strict inheritance* is allowed, i.e., specialisation by deleting any of the superclass's characteristics is prohibited. However, characteristics can be inherited from more than one superclass (*multiple inheritance*). The inheritance relationships (the *inheritance hierarchy*) resulting from the specialisation of MO classes can be visualised as graph (cf. Figure B.3 which shows the inheritance hierarchy defined by the *Definition of Management Information* [X.721]).

The MO class *top* is designated as the ultimate superclass in the OSI management inheritance hierarchy and every other MO class is a direct or indirect subclass of *top*. For example, the MO class *objectCreationRecord* in Figure B.3 is defined to be a direct subclass of *eventLogRecord* and so it is an indirect subclass of *logRecord* and also of *top*. Thus, in addition to its class-specific properties, *objectCreationRecord* will possess all the characteristics of each of the three superclasses.

An essential requirement in the context of systems management is to maintain interoperability between managing and managed systems when either the managed system is enhanced or one or more MO definitions are extended. With respect to a given MO it must be possible for a managing system to monitor and control another managed system of equal knowledge, another managed system of less knowledge, and another managed system of greater knowledge concerning the MO's class definition. In order to support this requirement, the concept of *allomorphism* has been introduced. Allomorphism is the ability of an MO that is an instance of a given class to be managed as an instance of one (or more) other MO class(es) (called *allomorphic class*). The two MO classes are not necessarily related by inheritance. The exact rules that ensure that an MO class is compatible with a second MO class are described in the *Management Information Model* [X.720].

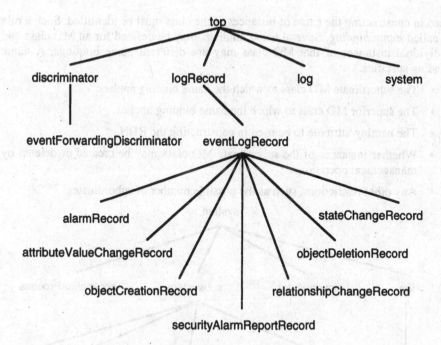

Figure B.3: Inheritance hierarchy of standardised managed object classes in X.721

B.2.3 Containment and Naming

To facilitate the naming of MOs, they are arranged in a hierarchical way by use of the *containment* relationship which may be used to express logical (a *layer entity* may contain *connections*) or physical (*equipment* may contain *circuits*) containment relations between the resources represented by the MOs. Each MO (termed the *subordinate* MO) is contained in exactly one other MO (termed the *superior* MO), and may itself contain an arbitrary number of MOs of the same or different MO classes. Containment and existence of MOs are closely related because an MO can only exist if its superior object exists. Note that containment is a relationship between MO instances, not classes (in contrast to the inheritance relationship).

Arranging MOs according to their containment relationships forms the naming tree. A superior MO is an unambiguous naming context for all of its subordinate MOs. The name of an MO which identifies it as being in the scope of the superior MO is called its *relative distinguished name* (RDN). The top level of the naming tree is referred to as *root* which is a null object. A *system* MO represents the managed system. The *local distinguished name* of a given MO specifies the name of the MO with respect to the *system* MO by concatenating relative distinguished names, starting from the *system* MO and continuing down the naming tree to end with the RDN of the given MO. The *global distinguished name* of this MO is constructed by concatenation of the identifying name of the *system* MO in the context of the global root and the local distinguished name of that MO.

For each instantiable MO class, i.e., an MO class which is defined not just for inheritance purposes, the naming attribute and the superior MO classes that may be used in constructing the name of instances of the class must be identified. Such a rule is called *name binding*. Several name bindings may be defined for an MO class and individual instances of that MO class may use different name bindings. A name binding specifies:

- The subordinate MO class to which the name binding applies.
- The superior MO class to which the name binding applies.
- The naming attribute to be used in constructing the RDN.
- Whether instances of the subordinate MO class may be created or deleted by management operations.
- Any other restrictions, such as the possible number of subordinates.

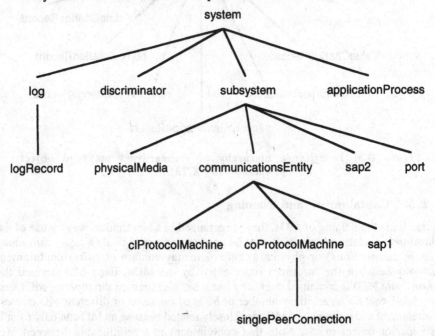

Figure B.4: Naming schema defined by OSI Systems Management in X.721

Every MO inherits from the *top* MO class the name binding attribute. This attribute contains the object identifier of the name binding which is in use between the MO and its superior. The collection of naming bindings is a naming schema. The naming tree of a given managed system is a specific instantiation of the overall naming schema. Figure B.4 shows the naming schema currently defined by the OSI standards *Definition of Management Information* [X.721] and *Generic Managed Information* [X.723].

B.2.4   Registration of Management Information

Management of a remote open system is based on the exchange of management information by means of management protocols. Within a management protocol, the management information is identified by the *object identifier* (OID) assigned to it during a registration process. An object identifier is an unambiguous name for a body of information called *information object*. The form of an object identifier is a sequence of object identifier components which are numeric values. The set of all object identifiers can be represented in a tree structure. This *object identifier tree* is a tree whose vertices correspond to administrative authorities responsible for allocating arcs from that vertex.

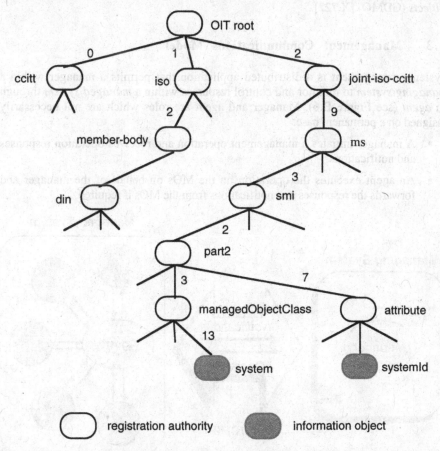

Figure B.5: Object identifier tree

Each arc of the tree is labelled by an object identifier component. Each information object to be identified is allocated precisely one vertex, normally a leaf, and no other information object is allocated to that same vertex. Thus an information object is uniquely and unambiguously identified by the sequence of numeric values, i.e., object identifier components labelling the arcs in a path from the root to the vertex which has been allocated to the information object. As defined in *Procedures for the operation of OSI Registration Authorities - Part 1: General procedures* [X.660] the responsibility for allocating unique object identifiers can be delegated through several *registration authorities*, each possessing an OID. Therefore, the non-leaf entries of the object identifier tree do not represent information objects themselves but registration authorities, while the leaves comprise the individual definitions.

The object identifier concept and the initial structure of the object identifier tree is described in *Specification of Abstract Syntax Notation One* [X.208]. The three nodes on the first level following the root are {ccitt}, {iso}, and {joint-iso-ccitt}. Standardised definitions in the context of OSI Systems Management are registered below {joint-iso-ccitt ms}. Figure B.5 depicts a small part of the overall registration tree for OSI Systems Management information objects. The description of the structure of this subtree is contained in *Guidelines for the Definition of Managed Objects* (GDMO) [X.722].

B.3   Management Communications Model

Systems management is a distributed application that permits a manager within a *managing system* to monitor and control resources within a *managed system* through an *agent* (see Figure B.6). Manager and agent are roles which are not necessarily assigned on a permanent base:

- A manager initiates a management operation and receives operation responses and notifications.

- An agent executes the operation on the MOs on behalf of the manager and forwards the responses and notifications from the MOs if required.

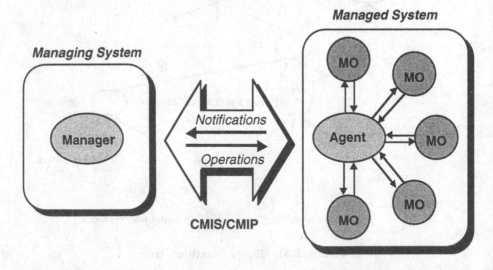

Figure B.6: Communications model of OSI Systems Management

Since the information model greatly influences the kind of management exchanges that may take place between a manager and an agent, the required generic management operations defined by CMIS and CMIP allow the manipulation of attributes and MOs and the transmission of actions and notifications.

An MO is selected by specifying its class and its instance name. A *scoping* parameter, which identifies *a base object* and a section of the containment tree below the base object, may be employed in some of the manager-initiated services to enable multiple objects to be selected for the performance of the management operation. In addition, *filtering* can be used to define assertions about attribute values of a scoped object. The operation will be performed only on MOs for which the specified filter evaluates to true. The following synchronisation conditions can be specified for an operation if it is to be executed on several MOs:

- *Atomic.* If the operation cannot be performed on all selected MOs, it is not performed on any MO, otherwise it is performed on all MOs.

- *Best effort.* The operation is executed on each selected MO independently of the failure of the operation on a particular MO.

A *Systems Management Application Entity* provides management functions (see section B.4) to manager and agents and makes use of the generic primitives of the *Common Management Information Service Element* (CMISE). The *Association Control Service Element* is used to establish, release, and abort management associations whereas the CMIS operations are conveyed by the *Remote Operations Service Element*.

B.4 Management Functional Model

In order to group the requirements on management capabilities, the OSI *Management Framework* [X.700] distinguishes five *Systems Management Functional Areas*:

- *Fault management* deals with the detection and correction of faults or abnormal operations of resources including maintenance and evaluation of error logs, fault tracing, and execution of diagnostic tests.

- *Configuration management* is concerned with the creation, modification, and deletion of MOs and with the reporting of configuration changes.

- *Accounting management* comprises the monitoring of access to the resources, collection and reporting of accounting information as well as the setting and modification of usage limits and tariffs.

- *Performance management* supports the monitoring and control of the performance of resources. This involves gathering statistical information on resource utilisation, deteriming system performance under normal and simulated conditions, and maintaining and evaluating system state histories.

- *Security management* is concerned with protecting the MOs. This includes the maintenance of security-relevant information such as access logs and the reporting of security-related events.

Based on these requirements, a set of *management functions* has been defined and documented in the parts of ISO 10164 / X.730ff. (cf. Table B.1). The specification of a management function can include one or more of the following:

- *Systems management services*, i.e., functionality that represents added value beyond that available from CMISE (or other application service elements) such as restrictions on the information content allowed or the temporal ordering of supporting service primitives.

- *Generic definitions* of MOs, attributes, actions, and notifications that address particular functional requirements.

- *Systems management functional units* which identify a specific set of systems management services and which may relate the use of these service to particular MO classes.

Each management function can add its specific management information to the common management information defined by CMIS, i.e., the management functions parameterise the CMIS services in order to provide the necessary detail.

ISO/IEC standard	Title	ITU-T Rec.
10164-1	Object Management Function	X.730
10164-2	State Management Function	X.731
10164-3	Attributes for Representing Relationships	X.732
10164-4	Alarm Reporting Functions	X.733
10164-5	Event Report Management Function	X.734
10164-6	Log Control Function	X.735
10164-7	Security Alarm Reporting Function	X.736
10164-8	Security Audit Trail Function	X.740
10164-9	Objects and Attributes for Access Control	X.741
10164-10	Usage Metering Function	X.742
10164-11	Metric Objects and Attributes	X.739
10164-12	Test Management Function	X.745
10164-13	Summarisation Function	X.738
10164-14	Confidence and Diagnostic Test Categories	X.737
10164-15	Scheduling Function	X.746
10164-16	Management Knowledge Management Function	X.750
10164-17	Changeover Function	X.751
10164-18	Software Management Function	X.744
10164-19	Management Domain and Management Policy Management Function	X.749
10164-20	Time Management Function	X.743
10164-21	Command Sequencer	X.753
10164-xx	Response Time Monitoring and Histogram Generation Function	X.748

Table B.1: Systems Management function standards

C Inter-Domain Management Information Service

In an open telecommunications environment, both network administrators and service users require an efficient and effective information service which provides details about services, operators, customers, networks, network elements, customer contracts, etc. Such requirements create a significant challenge and result in several issues that need to be tackled. The general problems relating to the openness and distributed nature of the envisaged service environment include the following:

- How to find the communication addresses of contact persons or operations systems for inter-domain management.

- Where to place more static / global (business management) information, for example, service advertisements.

- How to support distribution transparency of (management) services and related management information located on different management systems.

- How to perform operations on dependent managed objects distributed over several cooperating domains (i.e., how to locate the corresponding management systems).

In order to solve these problems, an information base is required that can store the management data relevant to cooperative management scenarios throughout the IBC environment. The data to be stored includes not only information on telecommunications resources that is contained in an operations system, or OSI Systems Management MIB [X.700], but also covers the organisational and operational information aspects of business, service, and network management. Such data is mainly stored in the X.500 directory, due to its more static nature [X.500]. This information is distributed over several domains and located in various end systems. The *Inter-Domain Management Information Service* (IDMIS) provides access to these different types of information via a single user interface. The database that is used by the IDMIS is called the *Inter-Domain Management Information Base* (IDMIB). This information base does not actually exist because the information is stored either in the X.500 directory or in OSI Systems Management MIBs.

The IDMIS is a very comprehensive and convenient service, somewhat similar to an enhanced directory service, but also enabling access to management information.

C.1 The Inter-Domain Management Information Model

The design of an information handling service to support the management of the IBC end-to-end services must support an integrated view of all kinds of management information wherever it is stored. Therefore, a common information model for the globally distributed management information had to be developed. The information model was constructed in a pragmatic manner, bearing in mind the benefits and characteristics of the principal components of the IDMIS: the X.500 directory and OSI Systems Management. X.500 and OSI Systems Management standards were designed to meet different objectives.

The *Directory Information Base* (DIB) stores fairly stable information intended primarily for location purposes as well as for supporting unambiguous and user-friendly naming. The DIB, with its global name space, is a suitable repository for storing and handling this information, which is represented by directory entries, or directory objects (DOs).

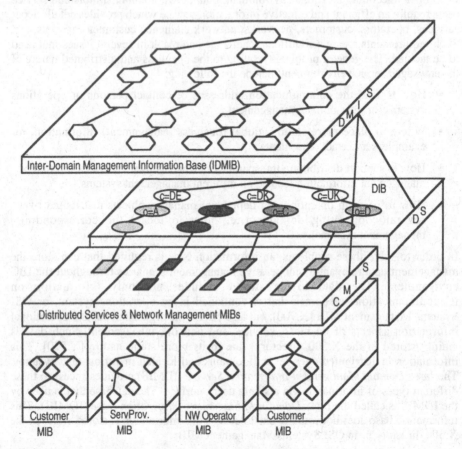

Inter-Domain Management Information Base (IDMIB)

Distributed Services & Network Management MIBs

Customer MIB   ServProv. MIB   NW Operator MIB   Customer MIB

Figure C.1: The integrated Inter-Domain Management Information Base (IDMIB)

Dynamic management information that can change very rapidly is suitable for representation as X.700 MOs and can be accessed or modified by OSI Systems Management services and protocols. This information is stored in local MIBs but may be required for end-to-end service management beyond a single end system. Additionally, network-related information may be of interest for end-to-end service management and may also need to be made available beyond a single end system.

In order to support end-to-end service management and achieve global accessibility, an integration of the two information models into one architecture with a unified name space was required. Fortunately, the information models are similar in both systems, so enabling a transparent integration by the IDMIB. Since the mechanisms for naming objects are comparable in both the DIB and the MIB, it was possible to establish a common name space that provides each object with a unique name and guarantees that it can be identified throughout the whole distributed environment. The IDMIS can therefore be used to access transparently all objects in the IDMIB, while hiding the nature and the location of objects related to inter-domain management. However, there are still two kinds of objects: DOs which have only attributes, and MOs which are described by attributes but which also exhibit some management behaviour as they can offer actions and emit notifications. These differences are hidden by the IDMIS so that the user does not need to be aware of the two separate systems, X.500 directory and OSI Systems Management.

In order to abstract from these differences, the concept of an Information Object (IO) was introduced. IOs represent the information aspects of objects made accessible by the IDMIS. They are merely an abstraction of MOs and DOs as seen from the perspective of the IDMIS user. In this sense, IOs are the objects which are visible to, and handled by, the IDMIS. An IO thus provides a unified view of the stored information and allows access to this object type via a similarly unified interface. The entire set of IOs, no matter whether MO or DO, constitutes the IDMIB as illustrated in Figure C.1. The IDMIB is hierarchically structured by X.500 structure rules in the upper levels, OSI Systems Management name bindings in the lower levels, and through the naming convention for integrating both systems which combines all local MIBs into one name space with the root of the Directory Information Base as the global root for all IOs. This naming convention has been standardised in the OSI Systems Management *Management Knowledge Management Function* [X.750].

C.2 Service Description

The central component of the IDMIS is an integrated inter-domain management information model which combines features of the models used in the X.500 directory and in OSI Systems Management. In this way, the design of the IDMIS is consistent with the standards. It provides a homogenous solution to a critical problem of inter-domain end-to-end service management, namely the efficient access to local and external information in a seamless manner. Consequently, the IDMIS enables the transparent administration and retrieval of end-to-end management information wherever it is stored, while hiding from the user the protocol used, whether the Directory Access Protocol (DAP) or the Common Management Information Protocol (CMIP).

The main areas where the IDMIS is applied are identification, location, and addressing. Unambiguous identification is achieved by assigning unique names to the individual information objects. Location problems can be solved by the comprehensive facilities of the IDMIS to search the IDMIB for the information required. Addressing is done by the IDMIS itself, provided that the information needed for addressing an end system (i.e., the presentation address) is stored in the IDMIB as well.

The IDMIS provides a uniform service interface to all information objects visible for inter-domain management. It provides the full capabilities of the X.500 directory service and the Common Management Information Service (CMIS). Additionally, it allows operations to be executed on information object groups distributed over the whole inter-domain management environment. For example, scoping and filtering can be performed on objects located in several customer MIBs, and change operations can be performed on directly related objects using transaction processing methods. The operations covered by the IDMIS can be classified according to the following three categories:

- *Information Retrieval*: reading, listing subordinates of specific objects within the IDMIB, searching, etc.
- *Information Control*: creating new objects, modifying existing objects, etc.
- *Event Notification*: functions that enable the reception of event reports which arrive asynchronously.

Provided Services	Services Used	
IDMIS	Directory Service	CMIS
Retrieval		
IDMIS-Read	Read	M-GET
IDMIS-Compare	Compare	M-GET
IDMIS-Search	Search	M-GET
IDMIS-List	List	M-GET
IDMIS-Cancel	Abandon	M-CANCEL-GET
Control		
IDMIS-Create	AddEntry	M-CREATE
IDMIS-Delete	RemoveEntry	M-DELETE
IDMIS-Change	ModifyEntry	M-SET
IDMIS-Action	n/a	M-ACTION
Notification		
IDMIS-InitReport	n/a	M-CREATE (Discriminator[1])
IDMIS-TerminateReport		
IDMIS-Report	n/a	M-DELETE
	n/a	M-EVENT-REPORT

Table C.1: Mapping from IDMIS services to the directory services and CMIS

Table C.1 shows the relationships between the IDMIS services provided at the IDMIS API and the standardised services used to access directory objects (via DAP) and managed objects (via CMIS/CMIP). Users are provided with an easy-to-handle service interface which clearly separates semantically distinct retrieval operations (read, list, search, compare). A detailed specification of the functionality offered by the IDMIS can be found in the RACE Common Functional Specification H.430 [CFS-H430].

---

[1]   IDMIS services at the Notification Port are the creation and deletion of Event Forwarding Discriminators (EFD) and the delivery of Event Reports.

C.3   The IDMIS Implementation Architecture

The PREPARE project validated this cooperative inter-domain communications architecture through the implementation of a communications management demonstrator on the PREPARE broadband network testbed. The realisation of an IDMIS was important to ensure that management application developers had a single integrated interface to the service that supported operations on information distributed throughout the testbed domains.

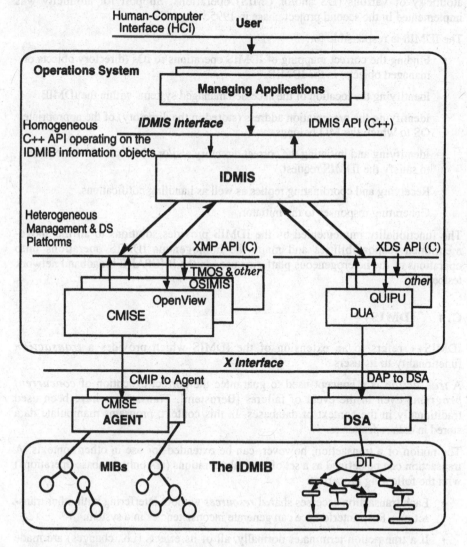

Figure C.2: The IDMIS implementation architecture

The management of the testbed domains was supported by a number of different platforms (including OpenView, OSIMIS, and TMOS). The IDMIS architecture provided application designers and implementers with a single object-oriented interface to both CMIS and Directory User Agent operations as shown in Figure C.2. Use of this interface facilitated the reuse and porting of management application code in heterogeneous environments.

The central component in this architecture is the IDMIS, which administers the knowledge about the structural organisation of the IOs and the tree structure of the IDMIB.

In certain cases (synchronisation atomic) the IDMIS may be required to guarantee the atomicity of various (DS and/or CMIS) operations. Support for atomicity was implemented in the second project phase in 1995.

The IDMIS is responsible for:

- Finding the correct mapping of IDMIS operations to IOs (directory objects or managed objects) in the IDMIB.
- Identifying the location of the accessed managed systems within the IDMIB.
- Identifying the presentation address (stored in the directory) of the appropriate OS to which the MO belongs.
- Identifying and initiating the correct directory and/or CMIS primitives required to satisfy the IDMIS request.
- Receiving and coordinating replies as well as handling notifications.
- Generating responses to the initiator.

The functionality implemented by the IDMIS provides solutions to all the above-mentioned responsibilities and maps the homogenous IDMIS operations onto operations of the heterogeneous platforms used in the PREPARE broadband network testbed.

C.4   IDMIS++

IDMIS++ refers to an extension of the IDMIS which provides a *transaction* functionality to its users.

A *transaction* is a concept used to guarantee the correct execution of *concurrent programs* even in the event of failures [Bernstein]. Transactions have been used traditionally in the context of databases. In this context, programs manipulate data stored in data repositories.

The notion of a transaction, however, can be extended for use in other contexts. A transaction can be defined as a set of general operations (not only database operations) with the following properties:

- Each transaction accesses shared *resources* without interfering with other transactions. Such interference can generate inconsistencies in a system.
- If a transaction terminates normally, all of its effects (i.e., changes) are made permanent; otherwise all of its effects are cancelled (i.e., the state of the resources is set to the original state prior to the start of the transaction).

Transactions are used in centralised and in distributed systems, where they provide a way to guarantee that the state of one or more systems is modified consistently. In the context of PREPARE, the IDMIB represents a shared resource that is accessed by multiple IDMIS users. Concurrent accesses to the IDMIB or the occurrence of failures can lead to inconsistent states in the IDMIB.

IDMIS++ extends the IDMIS interface with operations to define and control transactions. Through these operations an IDMIS++ user can group a set of accesses to the IDMIB as a transaction. IDMIS++ will execute this set of accesses taking into consideration the occurrence of failures and concurrency on shared data. The transaction support provided by IDMIS++ is a concept to control operations that span the directory and a set of managed systems. Full consistency maintenance, however, can only be provided for managed systems. Due to the lack of a suitable transaction support by the X.500 directory, transactions that access the directory can only be used in a restricted way.

C.4.1 ACID Properties

To describe the characteristics of a type of transaction, the *Atomicity*, *Consistency*, *Isolation*, and *Durability* properties are frequently used. These properties are referred as *ACID properties*:

- **(A)tomicity:** either all of the operations of a transaction are executed or none of them is.

- **(C)onsistency:** the operations of a transaction taken as a group do not generate an inconsistent state. Which states are inconsistent depends on application semantics.

- **(I)solation:** until the completion of a transaction, the result is not *visible* to other operations that do not take part in this transaction.

- **(D)urability:** the effects of the execution of a transaction are permanently maintained by the system, even in the event of failures.

These properties provide sufficient conditions for the correctness of the execution of transactions in a system. If each transaction execution in a system obeys these properties, the consistency of the system is always preserved.

C.4.2 The Structure of an IDMIS++ Transaction

An IDMIS++ transaction is composed of operations to manipulate IOs in the IDMIB and of operations to control the transaction. Figure C.3 shows an example of the submission of a hypothetical IDMIS++ transaction. In this example, the operations on the IDMIB are mapped by IDMIS++ onto directory services and onto management services to access MIBs of two different managed systems (managed systems 1 and 2 in Figure C.3).

The operations available to manipulate the IDMIB are the ones provided by the IDMIS (see Table C.1). Additional operations to control transactions are provided by the IDMIS++. They are shown in Table C.2 with a brief description of their functionality. Conceptually they are provided at the control port of the IDMIS.

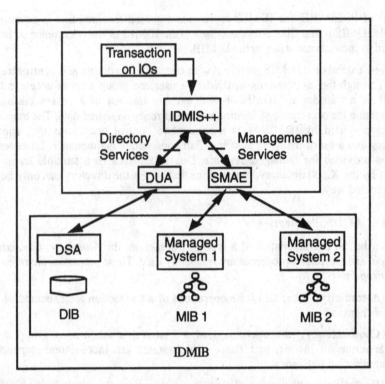

Figure C.3: Example of the submission of transactions to IDMIS++

To submit a transaction to IDMIS++, the IDMIS++ user first calls the *IDMIS_begin_transaction* operation to indicate the start of a transaction. Afterwards, the user calls IDMIS++ operations to manipulate the IDMIB. Each of these operations takes part in the transaction. After all operations are submitted, the IDMIS++ user terminates the transaction by requesting either *commitment* or *rollback*. Committing a transaction means that no failures were detected during the transaction and that the state of all systems involved is consistent. Therefore, the effects of the transaction will be made permanent in all systems that took part in it. If a transaction cannot be committed it has to be rolled back and all effects of the transaction will be cancelled. For commitment, the IDMIS++ user calls the *IDMIS_commit* operation. For rollback, the *IDMIS_rollback* operation has to be called. The decision to either rollback or commit a transaction is made by the IDMIS++ user[1].

The relationship between the systems that take part in an IDMIS++ transaction can be represented in the form of a tree of depth one. A tree that represents an IDMIS++ transaction is called a *transaction tree*. Figure C.4 depicts the transaction tree for the transaction example of Figure C.3.

The *nodes* of a transaction tree are the systems where the transaction is performed. In Figure C.4 there are four nodes:

- The system where IDMIS++ runs.

---

1   Failures, however, can unilaterally cause a transaction to be rolled back.

- The directory.
- The managed systems.

Service	Functionality
IDMIS_begin_transaction	Marks the beginning of a transaction
IDMIS_rollback	Requests the cancellation of the effects of a transaction
IDMIS_commit	Requests that the effects of the transaction are made permanent

Table C.2: IDMIS++ services for transaction control

The root of the tree is called the *coordinator*. The other nodes are called *subordinates*. In a transaction tree, the system where IDMIS++ runs will always be the coordinator. A subordinate can either be the directory or a managed system. Coordinator and subordinate in a transaction tree are connected by a *transaction branch*.

During the processing of the operations in a transaction, the coordinator (IDMIS++) identifies the subordinates (directory or managed systems) where the operations have to be executed. A transaction branch is set up to each subordinate if no branch yet exists.

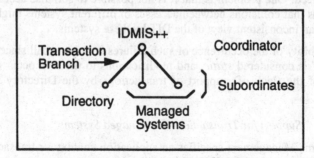

Figure C.4: Example of a transaction tree

When the IDMIS++ user requests the termination of a transaction, the coordinator must coordinate the activities on all transaction branches in order to come to the desired transaction outcome (*commitment* or *rollback*) over the whole transaction tree.

C.4.3 Implementation Aspects

IDMIS++ has to synchronise the operations in the directory and in the managed systems involved in a transaction to attain a consistent outcome (commitment or rollback) of the transaction. Due to the scope of the project involving more than only transaction support, the use of the ISO/ITU-T Transaction Processing (TP) protocol [X.860][X.861][X.862] and the Commitment, Concurrency and Recovery (CCR) protocol [X.851][X.852] was not encompassed in the implementation of IDMIS++, although the approach taken implements a variation of these protocols. The IDMIS++ uses only X.500 directory and OSI Systems Management services to provide its own services including the transaction support.

C.4.3.1 Support for Transactions on the Directory

The X.500 directory provides no support for the execution of transactions [Richter]. The provision of such support requires more functionality than actually provided by the directory and the IDMIS++ transaction support does not consider this provision. The execution of parts of a transaction by the directory, therefore, are supported only in a limited way and are entirely controlled by the coordinator (IDMIS++).

In order to provide atomic execution of directory services, the coordinator must guarantee that the data in the directory is in a consistent state, even in an event of failures. The coordinator can provide this guarantee in a certain way because the syntax and the semantics of all X.500 directory operations are well defined. Having knowledge of the effects that an X.500 directory operation has on the Directory Information Tree (DIT), the coordinator can follow a specific strategy to maintain the consistency of the DIT. The coordinator tries to execute the operations of a transaction branch on the directory. In case of a failure, the coordinator restores the state of the DIT to the original state prior to the first operation that was executed within the transaction branch.

For operations that only read values of the DIT, no action at rollback has to be carried out since the state of the DIT has not changed. For each operation that changes the state of the DIT, specific *compensating* operations [Richter] can be performed by the coordinator in order to restore the original state. Since there is no transaction support provided by the directory, the re-establishment of the original state of the DIT cannot be guaranteed. One problem is that it is not possible to serialise accesses to the DIT. This means that collisions between accesses of different systems might occur, which can cause an inconsistent view of the DIT by these systems.

The probability of the occurrence of such failures is rather small since the data stored in the DIT is considered *static*, and modifications to the DIT occur only rarely. An analysis of the (lack of) support of transactions by the Directory is presented in [Richter].

C.4.3.2 Support for Transactions on Managed Systems

OSI Systems Management specifies an application context for transaction support of management operations [X.702]. This application context was not considered because it requires the use of the TP protocol. The functionality described in these standards, however, allows the development of a suitable support for transaction processing in managed systems. The CMISE services are used to support the necessary communication between the systems.

Figure C.5 shows the components considered at a managed system to support IDMIS++ transactions. The control of transaction branches over managed systems relies on a transaction support by the managed systems themselves. A special MO must exist in each managed system to support transactions. This object, called *TCMO* (*Transaction Control Managed Object*), provides operations that are invoked by the coordinator in order to control the transaction branch to this managed system.

The operations provided by the TCMO are invoked by the coordinator by using the *M-Action* service of CMISE. These operations support the dialogue between the TCMO and the coordinator to start and stop transactions. On each association to a managed system only one transaction branch may exist at a time. A transaction branch is set up for an association when the coordinator invokes a specific operation of the TCMO.

From this point on up to the termination of the transaction (either through commitment or rollback), every management operation on any MO that is executed over this association is considered to belong to the transaction branch. To request commitment or rollback of the transaction, other TCMO operations are invoked. The protocol between the coordinator and the TCMO is an implementation of the *2-Phase commit protocol with presumed rollback* [Bernstein].

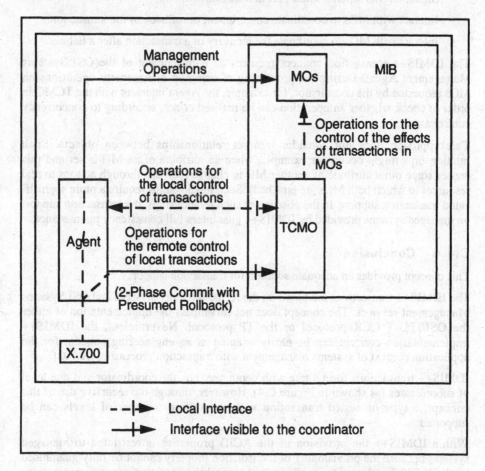

Figure C.5: General structure of managed systems

Because much flexibility is possible when designing an MO, for example in the definition of actions and in the various ways they can interact with real resources, only the MO itself can determine what constitutes its states (and the states of the real resource, if there is one attached to the MO). For the control of rollback and recovery it is necessary that the MOs can be set to specific states (*initial state* in the case of rollback or *final state* in the case of commitment). Therefore, the IDMIS++ transaction concept relies on a support of transactions by the MOs as well. Only MOs that provide transaction support can take part in a transaction. If an MO does not support transactions it may only take part in the transaction if the state of the MO remains unchanged.

Locally at the managed systems, the TCMO coordinates the execution of transactions over other objects of the MIB. Each MO that provides transaction support must provide a local interface to the TCMO. The operations of this interface are invoked locally by the TCMO and are not visible to the coordinator. Through the operations at this interface, the TCMO can:

- Signal an MO when it takes part in a transaction.

- Interact with MOs to coordinate commitment or rollback of the transaction.

- Interact with MOs to coordinate the recovery of a transaction after a failure.

The IDMIS++ transaction concept requires an enhancement of the OSI Systems Management Agent. During the processing of common management operations on MOs requested by the coordinator, for example, the Agent interacts with the TCMO in order to check whether an operation can be realised or not, according to concurrency control aspects.

This approach does not consider indirect relationships between objects. Such relationships might occur, for example, when an attribute of an MO is set and this causes some other attribute of another MO to change (perhaps through actions to real resources to which both MOs are attached). Such relationships require a more sophisticated transaction support. In the absence of such relationships, the transaction support on managed systems provided by IDMIS++ guarantees full consistency maintenance.

C.4.4   Conclusion

This concept provides an adequate solution for transaction support.

The IDMIS++ transactions are based on the X.500 directory services and OSI Systems Management services. The concept does not encompass the implementation of either the OSI/ITU-T CCR protocol or the TP protocol. Nevertheless, the IDMIS++ implementation concept can be easily adapted as an engineering concept for the application context of systems management with transaction processing [X.702].

IDMIS++ transactions form a tree with depth one, i.e., the coordinator and one level of subordinates (as shown in Figure C.4). However, through the recursive use of this concept, a type of nested transaction with an arbitrary number of levels can be supported.

Within IDMIS++ the provision of the ACID properties is restricted to managed systems because the provisioning of the isolation property cannot be fully guaranteed by the X.500 directory. Due to the lack of a suitable transaction support, transactions that access the directory can only be supported in a restricted way. Therefore, the concept does not guarantee the absence of conflicts in all cases.

This concept has to be further developed in order to provide support for transactions in cases where the autonomy properties of systems have to be taken into account. Some approaches are currently under study to support transactions in environments of autonomous systems, for example, the use of S-transactions [EliaVeij]. Indirect relationships between MOs are not considered in this concept either. IDMIS++ transactions were a first step towards attaining a suitable transaction support for the whole PREPARE multi-domain environment.

C.5 Advantages Offered by IDMIS / IDMIS++

The global information architecture establishes an integrated Inter-Domain Management Information Base. It provides a common naming scheme, rules, and conventions for supporting the structuring of information objects and for allowing access to service-related information within the IBC environment.

The advantages of using the IDMIS to access information required to support end-to-end service management in heterogeneous environments are as follows:

- A conceptual integration of the X.500 DIB and the OSI Systems Management MIBs is realised by virtually linking their name spaces.

- The information objects (DOs and MOs) are identified by unique names within a universal, common information structure.

- The physical distribution and location of management information does not have to concern the application developer and service user. It is the responsibility of the IDMIS to locate the correct information from the context of the query.

- The IDMIS provides access transparency throughout the inter-domain environment, i.e., the user only has to know the name of an object and not where it is stored.

- Specialised easy-to-use retrieval functions (e.g., a search operation involving several domains) can be offered to the service user.

- Information retrieval and updating can be performed via a unified interface. Also, the user (human manager or management application) is relieved from dealing with details of the underlying systems, such as how a connection is to be established or which special control parameters for directory queries have to be used (e.g., *prefer chaining, do not use copy*, etc.).

- Control functions to modify the distributed IDMIB are supported, i.e., to insert new IOs into it, to remove information from it, to execute actions of IOs, and to modify existing IOs. With IDMIS++ the concurrent execution of such functions can be done correctly even in case of failures (with the limitations described in section C.4.4).

- The IDMIS provides cache mechanisms to avoid unnecessary access to the Directory System, which can cause performance problems.

In addition, the definition of an object-oriented application programming interface (IDMIS API) and the features of the IDMIS have the following benefits:

- The use of object-oriented software design and development methods is supported by the C++ IDMIS API.

- Management Application modules can be reused more easily on different hardware platforms.

A summary of the functionality offered by the IDMIS leads to the conclusion that it is a very comprehensive and convenient service which enables the transparent administration and retrieval of inter-domain management information wherever it is stored.

The object-oriented integrated IDMIS interface simplifies the use of advanced information processing techniques for the development of management applications and enables the reuse of management application software modules in heterogeneous management platforms. The support of portability and interworking for management applications from different vendors will play a major role in the future open service market. The concepts of the IDMIB and the IDMIS can be seen as one step in this direction.

D TMN-Based Inter-Domain Management of IN

D.1   Introduction

The main goal of PREPARE was to consider TMN-based management of services in the context of an open service market resulting from the implementation of the Open Network Provision (ONP) Directive from the European Commission. Although PREPARE was not a project concerned with Intelligent Network (IN) services as defined by the International Telecommunication Union (ITU-T) in the Q.1200 series, we recognised that IN will be one of the major telecommunications markets at the beginning of the 21st century. Therefore, we found it useful to apply our experience of management to IN, considering TMN-based management of IN networks and services in the context of an open service market. The result of this work is presented in this appendix.

Section D.2 presents the IN functional and physical entities considered in this appendix while section D.3 depicts potential configurations of IN functional and physical entities when an open service market is envisaged.

Section D.4 describes our views about TMN-based management of an IN environment where network and services are provided by a single operator. Section D.5 expands these views by considering a situation where the network and the IN services are provided by independent organisations.

D.2   IN Network and Services Provided by a Network Operator

As a starting point, Figure D.1 represents the IN functional and physical entities defined in *Intelligent Network Distributed Functional Plane Architecture* [Q.1204], and *Intelligent Network Physical Plane Architecture* [Q.1205], which are considered in the context of this appendix. Service Management Function (SMF), Service Management Access Function (SMAF) and Service Management Point (SMP) have been replaced by TMN functional and physical entities as suggested by the *Baseline Document on the Integration of IN and TMN* from ETSI [NA43308]. It should be noted that this appendix considers only the IN service processing phase. The Service Creation Environment Function (SCEF) and the Service Creation Environment Point (SCEP) are not represented. Additionally, Figure D.1 indicates a possible mapping of IN functional entities onto physical ones (functional entities in italics in Figure D.1 are optional in the context of a given physical entity).

The IN functional model represents entities serving different purposes: entities related to the provision of bearer services (e.g., fixed or mobile telephone services) and entities dedicated to the provision of "intelligent" IN services. The former group contains functional entities which are the representation in the IN environment of network elements providing bearer services: Call Control Agent Function (CCAF) and Call Control Function (CCF). These elements provide their services on their own and are independent of the IN technology. The latter group contains functional entities involved in the provision of "intelligent" IN services i.e., providing the value added by the IN technology. This group contains the Service Switching Function (SSF),

Specialised Resource Function (SRF), Service Control Function (SCF), and Service Data Function (SDF).

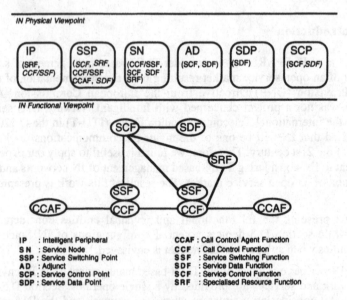

Figure D.1: IN functional and physical entities

D.3   IN Network and Services in an Open Service Market

In the context of an open service market, independent service providers will be able to offer their services using an infrastructure provided by traditional network operators. Applied to the IN environment, this policy will favour the emergence of independent IN service providers, offering the "intelligent" part of IN services and leaving provision of the underlying bearer services to the network operators (which may also offer their own IN services). Figure D.2 depicts potential configurations of IN functional and physical entities in such a context.

The following assumptions apply to Figure D.2:

- CCAF, CCF and SSF belong only to the network operator domain while SCF, SDF, SRF belong to either the network operator domain or the independent IN service provider domain.

- Intelligent Peripheral (IP) without its CCF/SSF option, SDP and SCP can be instantiated in both the network operator and the independent IN service provider domains to support SRF, SDF and SCF.

- Adjuncts (ADs) which are closely connected to CCF/SSF cannot be instantiated in the independent IN service provider domain.

Actual multi-provider configurations will instantiate a subset of this overall potential configuration (e.g., with the SCF and SDF in the independent IN service provider domain and the SRF in the network operator domain).

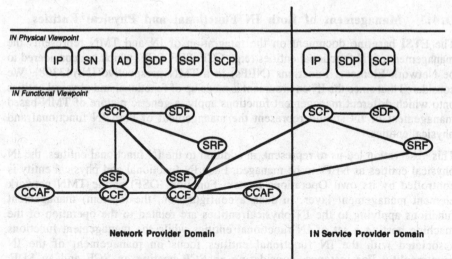

Figure D.2: IN functional and physical entities in a multi-provider configuration

D.4   TMN-Based Management of IN Network and Services Provided by a Network Operator

Based on the observations in previous sections, we introduce Figure D.4 which reflects the PREPARE interpretation of TMN-based management of IN networks and services when they are both provided by a single organisation (i.e., a network operator). The main differences compared with the mapping of IN service processing functional entities onto the TMN functional architecture proposed in the ETSI baseline document on the integration of IN and TMN [NA43308] (cf. Figure D.3) are justified below.

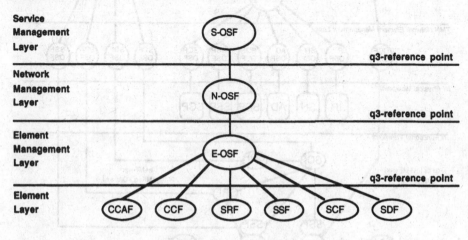

Figure D.3: Mapping of the IN service processing functional entities onto the TMN functional architecture

D.4.1 Management of Both IN Functional and Physical Entities

The ETSI baseline document on the integration of IN and TMN represents the management of IN functional entities (e.g., CCF, SSF, SCF) which are considered to be Network Elements Functions (NEFs) in a TMN perspective [NA43308]. We considered that since the IN environment is made up of functional and physical entities onto which different management functions apply, a generic picture of TMN-based management of IN should represent the management of both IN functional and physical entities.

This observation led us to represent, in addition to the IN functional entities, the IN physical entities as NEFs to be managed. Each IN functional and physical entity is controlled by its own Operations System Function (OSF) at the TMN network element management layer. In such a configuration, the (system) management functions applying to the IN physical entities are related to the operation of the machine hosting a set of IN functional entities while the management functions associated with the IN functional entities focus on management of the IN functionality. For instance, considering an SCP hosting an SCF and an SDF, hardware fault management is an SCP concern handled by the SCP OSF (in cooperation with other OSFs), while the management of data related to a Private Numbering Plan (PNP) in the context of an IN Virtual Private Network (VPN) service is an SDF concern to be handled by an SDF OSF (in cooperation with other OSFs).

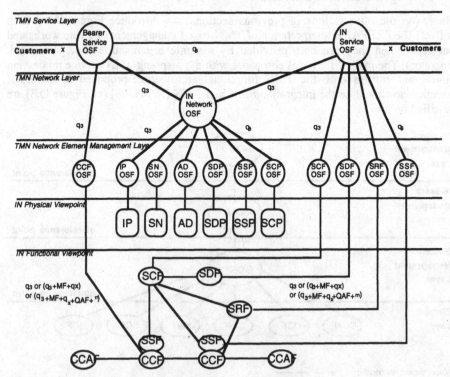

Figure D.4: **TMN-based management of IN networks and service provided by a network operator**

D.4.2   Bearer and Intelligent Service OSFs

As stated earlier, an IN structured network provides (plain) bearer services and richer "intelligent" services. Basic services can be sold and are meaningful on their own (e.g., fixed telephone services) and IN services can be sold as self-contained service offerings through their own commercial interface (e.g., a voice VPN service providing a PNP) even though dependent on underlying bearer services,

We reflected this *functional autonomy* of bearer to IN services by means of two service OSFs controlling the IN functional entities and the network element management layer OSF associated with the provision of bearer and "intelligent" services respectively.

Since the IN services depend on an underlying bearer service from which management information may be required or onto which management actions may be applied (depending on the nature of the IN service), it is necessary that the bearer and the IN service OSFs are able to exchange management information through a management interface.

D.4.3   Inter-OSF Relationships

Figure D.4 depicts an arrangement of OSFs that differs from the classic one proposed in *Principles for a Telecommunications Management Network* [M.3010] and considered in the baseline document [NA43308], where a service layer OSF interacts only with a network layer OSF. The following reasons justify our choice for setting up direct interfaces between service layer OSFs and network element OSFs controlling the IN functional entities:

- Though termed and represented as *network elements* in order to comply with the TMN architecture and terminology, the IN functional entities are service resources (service database, voice servers, etc.) rather than network components[1] (in the usual meaning of the term). For that reason, we did not feel it necessary to subordinate all management actions applying to IN functional entities to both a service layer and a network layer OSF.

- Originally, the IN architecture was designed for making the service control functions independent of the bearer network and this intention is reflected in the structure of the IN functional architecture. Therefore, it is sensible to apply the same principle to the management of an IN structured environment, avoiding any unnecessary dependency of the service management layer on the network management layer.

Nevertheless, management information exchange is still required between the network layer and service layer OSFs in order to give to the service layer OSF knowledge about the assignment of functional to physical entities, the configuration and traffic in the network, and status and fault management information related to the physical entities in the network. The management information exchange supporting these functionalities is supported by the management interfaces linking the network OSF to the bearer and the IN service OSFs.

---

[1]   This could be discussed in the case of the CCF/SSF functional entities which are closely related to particular pieces of network equipment.

D.5 TMN-Based Management of IN Network and Services in an Open Service Market

In the future open service market, several service providers will compete to provide their services on top of a single network infrastructure. This situation will lead to the need for opening the network operator management infrastructure to independent IN service providers and to allow a number of inter-organisation service management relationships to take place, such as the billing of the independent service provider. On the other hand, the network operator must protect its network integrity at all costs and, therefore, should filter management information flows crossing the inter-domain management interface.

In order to cater for these requirements, we introduce at the service management layer an *Interconnecting Service OSF* in charge of ensuring a mediation function (encompassing all aspects of an inter-organisation relationship) between independent IN service providers and a network operator. The interfaces supporting the inter-organisation relationships between the Interconnecting Service OSF and the OSF in the independent IN service provider domains are TMN X interfaces. Some important aspects to be covered by the Interconnecting Service OSF include :

- Access to basic service management functions by the service providers.
- Billing of the service providers.
- Protection of network integrity.
- Security (e.g., access).

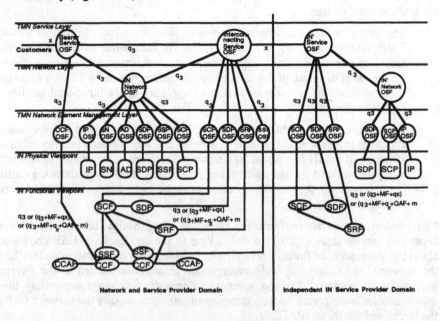

Figure D.5: TMN-based management of IN networks and services in an open service market

D.6 Conclusion

Taking the work carried out in ETSI on IN/TMN integration as a starting point, this appendix has outlined an architectural proposal for the management of IN structured networks and services in the context of an open service market where several organisations will cooperate and/or compete for the provision of intelligent telecommunication services to their customers.

The key points proposed are the identification of different entities (i.e., OSF in the TMN terminology) involved in the management of such a network and service environment as well as the introduction of an Interconnecting Service OSF in charge of ensuring safe and efficient *management interconnection* between a network operator and independent IN service providers.

Practical experiments based on the proposed solution need to be carried out in order to validate its adequacy and usefulness for solving the management issues raised by TMN-based management of IN networks and services for the open service market.

E Examples of PREPARE Information Models at the Service Layer X Interface

The appendix provides examples of PREPARE information models used for managing services described in chapter 4. The first section contains a simple version of the GDMO descriptions of the X interface for ATM service layer and the second section provides the GDMO definition of the PREPARE VPN service.

E.1 Public Network Service Layer X Interface

E.1.1 Managed Object Classes

xSystem MANAGED OBJECT CLASS
DERIVED FROM "Recommendation X.721 : 1992": system;
CHARACTERIZED BY
"Recommendation X.721 : 1992": administrativeStatePackage,
xSystemPackage PACKAGE
BEHAVIOUR xSystemBehaviour
ATTRIBUTES
        "CCITT Rec M.3100": alarmStatus GET,
        "Recommendation X.721 : 1992": administrativeStatePackage GET-
REPLACE,
NOTIFICATIONS
        "Recommendation X.721 : 1992": stateChange;;;
REGISTERED AS {xObjectClass 1};

uni MANAGED OBJECT CLASS
DERIVED FROM "Recommendation X.721 : 1992": top;
CHARACTERIZED BY
uniPackage PACKAGE
BEHAVIOUR uniBehaviour
ATTRIBUTES
        uniId GET,
        "Recommendation X.721 : 1992": administrativeState GET-REPLACE,
        "Recommendation X.721 : 1992": operationalState GET,
        "Recommendation X.721 : 1992": usageState GET;
NOTIFICATIONS
        "Recommendation X.721 : 1992": stateChange;;;
CONDITIONAL PACKAGES
        "Recommendation M.3100" tmnCommunicationsAlarmInformationPackages
        PRESENT IF "to be defined";
REGISTERED AS {xObjectClass 2};

e164Addr MANAGED OBJECT CLASS
DERIVED FROM "Recommendation X.721 : 1992": top;
CHARACTERIZED BY
e164AdrPackage PACKAGE
BEHAVIOUR uniBehaviour

ATTRIBUTES
       e164AdrNumber GET,
       "Recommendation X.721 : 1992": administrativeState GET-REPLACE,
       "Recommendation X.721 : 1992": operationalState GET,
       "Recommendation X.721 : 1992": usageState GET;
NOTIFICATIONS
       "Recommendation X.721 : 1992": stateChange;;;
REGISTERED AS {xObjectClass 3};

atmAccess MANAGED OBJECT CLASS
DERIVED FROM "Recommendation X.721 : 1992": top;
CHARACTERIZED BY
atmAccessPackage PACKAGE
BEHAVIOUR atmAccessBehaviour
ATTRIBUTES
       atmAccessPointId GET,
       "Recommendation X.721 : 1992": administrativeState GET-REPLACE,
       "Recommendation X.721 : 1992": operationalState GET
       freeBandwidth GET;
NOTIFICATIONS
       "Recommendation X.721 : 1992": stateChange;;;
CONDITIONAL PACKAGES
       "Recommendation M.3100" tmnCommunicationsAlarmInformationPackages
       PRESENT IF "to be defined";
REGISTERED AS {xObjectClass 4};

vpCP MANAGED OBJECT CLASS
DERIVED FROM "Recommendation X.721 : 1992": top;
CHARACTERIZED BY
vpCPPackage PACKAGE
BEHAVIOUR vpCPBehaviour;
ATTRIBUTES
       vpi GET,
       vpLinePointer GET,
       "Recommendation X.721 : 1992": administrativeState GET-REPLACE,
       "Recommendation X.721 : 1992": operationalState GET;
NOTIFICATIONS
       "Recommendation X.721 : 1992": stateChange;;;
CONDITIONAL PACKAGES
       "Recommendation M.3100" tmnCommunicationsAlarmInformationPackages
       PRESENT IF "to be defined";
REGISTERED AS {xObjectClass 5};

vpLine MANAGED OBJECT CLASS
DERIVED FROM "Recommendation X.721 : 1992": top;
CHARACTERIZED BY
vpLinePackage PACKAGE
BEHAVIOUR vpLineBehaviour
ATTRIBUTES
       vpLineId GET,

        a-E164Address GET,
        a-Pointer GET,
        z-E164Address GET,
        z-Pointer GET,
        azBandwidthDescriptor GET-REPLACE,
        zaBandwidthDescriptor GET-REPLACE,
        "Recommendation X.721 : 1992": administrativeState GET-REPLACE,
        "Recommendation X.721 : 1992": operationalState GET;
NOTIFICATIONS
        "Recommendation X.721 : 1992": stateChange;
        "Recommendation X.721 : 1992": attributeValueChange;;;
CONDITIONAL PACKAGES
        callInformationPackages PRESENT IF "to be defined",
        originatorInformationPackages PRESENT IF "to be defined",
        globalIdentifierPackages PRESENT IF "to be defined",
        "Recommendation M.3100" tmnCommunicationsAlarmInformationPackages
        PRESENT IF "to be defined";
REGISTERED AS {xObjectClass 6};

vpProfile MANAGED OBJECT CLASS
DERIVED FROM "Recommendation X.721 : 1992": top;
CHARACTERIZED BY
vpProfilePackage PACKAGE
BEHAVIOUR vpLineBehaviour
ATTRIBUTES
        vpProfileId GET,
        backGroundTraffic GET-REPLACE
        specialTrafficPeriodes GET, ADD-REMOVE;
        "Recommendation X.721 : 1992": administrativeState GET-REPLACE,
        "Recommendation X.721 : 1992": operationalState GET;
NOTIFICATIONS
        "Recommendation X.721 : 1992": stateChange;;;
REGISTERED AS {xObjectClass 7};

E.1.2   ASN.1 Defined Types Module

ASN1DefinedTypesModule {iso(1) member-body(2) denmark(202) ktas(101)
xServiceLayerPrepare(5) informationModel(0) asn1Modules(2)
asn1DefinedTypesModule(0)}
DEFINITIONS IMPLICIT TAGS ::=
BEGIN
--EXPORTS everything
IMPORTS
SimpleNameType FROM Attribute-ASN1Module{joint-iso-ccitt ms(9) smi(3)
part2(2) asn1Module(2) 1}
Pointer FROM M.3100, ASN1DefinedTypesModule{ccitt(0) recommendation(0)
m(13) gnm(3100) informationModel(0) asn1Modules(2)}

TrafficDescriptor FROM ETS52210-ASN1TypeModule { ccitt(0) identified-
organization(4) etsi(0) na52210(52210) informationModel(0) asn1Module(2)
asn1TypesModuleASN1(2)};
xInformationModel OBJECT IDENTIFIER ::= {iso(1) member-body(2) denmark(202)
ktas(101) xServiceLayerPrepare(5) informationModel(0)}
xObjectClass OBJECT IDENTIFIER ::= {xInformationModel
managedObjectClass(3)}
xParameter OBJECT IDENTIFIER ::= {xInformationModel parameter(5)}
xNameBinding OBJECT IDENTIFIER ::= {xInformationModel nameBinding(6)}
xAttribute OBJECT IDENTIFIER ::= {xInformationModel attribute(7)}

A-Pointer ::= Pointer

BackgroundTraffic ::= BandwidthDescriptor

BandwidthDescriptor ::= SEQUENCE {
        peakKB   [1]   INTEGER   OPTIONAL,
        peakCellRate   [2]   INTEGER   OPTIONAL,
        meanKB   [3]   INTEGER   OPTIONAL,
        meanCellRate   [4]   INTEGER   OPTIONAL,
        policingParameter   [5]   TrafficDescriptor OPTIONAL}

Digit ::= INTEGER(0..9)

E164addrNumber ::= SEQUENCE SIZE(15) OF Digit

FreeBandwidth ::= BandwidthDescriptor

PeriodType ::= ENUMERATED {
        absolute(0),
        month(1)
        week(2)
        day(3)
        hour(4)}

SpecialTrafficPeriods ::= SET OF TrafficPeriod

Time ::= SEQUENCE {
        year   [1] INTEGER   OPTIONAL,
        month   [2] INTEGER   OPTIONAL,
        day   [3] INTEGER   OPTIONAL,
        hour   [4] INTEGER   OPTIONAL,
        minute [5] INTEGER}

TrafficPeriod ::= SEQUENCE {
        period   PeriodType
        startTime   Time,
        stopTime   Time,

traffic   BandwidthDescriptor}

VpCreateProblemParameter ::= ENUMERATED {
        ok(0)}

Vpi ::= INTEGER(0..255)

Z-Pointer ::= Pointer

END

E.2   VPN Service Layer X Interface

The information model is split into two parts:

1. The general VPN management information model which specifies the technology independent aspects of the model.

2. The ATM Constant Bit Rate (CBR) service-specific VPN management information model which specifies the extension to the general information model designed and implemented in the PREPARE project to demonstrate the application of the VPN management service to an ATM network providing a Constant Bit Rate Virtual Path (VP) service.

E.2.1   General VPN Management Information Model

E.2.1.1   MO Schema Definitions

"Prep-Ext Vpn":customerPath MANAGED OBJECT CLASS
DERIVED FROM "Prep-Ext Vpn":vpnLink;
CHARACTERIZED BY
        customerPathPkg PACKAGE
BEHAVIOUR
        customerPathBeh BEHAVIOUR
 DEFINED AS
"The customerPath is a link that represents the end-to-end resource reservation requirements of the VPN customer. It is terminated by endPoints representing the edges of the customer's domain of interest.";;;;
REGISTERED AS {vpnMO 1};

"Prep-Ext Vpn":connectionDescriptor MANAGED OBJECT CLASS
DERIVED FROM "Prep-Ext Vpn":top;
CHARACTERIZED BY
connectionDescriptorPkg PACKAGE
BEHAVIOUR
 connectionDescriptorBeh BEHAVIOUR
 DEFINED AS
"A connectionDescriptor provides interface and routing information for a point.
userStreamSupported is the user stream for which this information is provided.
localEndPointList are the end points in the local network domain whose connection

this connectionDescriptor is supporting. remoteEndPointList are the end points in a
remote domain whose connection this connectionDescriptor is supporting.
networkLinkUsed is the network link in the local domain from which resources for the
userStream should be allocated.";;
ATTRIBUTES
connectionDescriptorId   GET,
userStreamSupported   GET-REPLACE,
localEndPointList   GET-REPLACE ADD-REMOVE,
remoteEndPointList   GET-REPLACE ADD-REMOVE,
networkLinkUsed   GET-REPLACE;;;
REGISTERED AS {vpnMO 2};

"Prep-Ext Vpn":endPoint MANAGED OBJECT CLASS
DERIVED FROM "ISO/IEC 10165-2":top;
CHARACTERIZED BY
pointPkg PACKAGE
BEHAVIOUR
 pointBeh BEHAVIOUR
DEFINED AS
"The endPoint represents the points at the end of a link. NetworkNode provides a
reference to an MO that represents the physical network node onto which the point
may be mapped. The point cannot be used if the administrativeState is Locked. The
operationalState indicates whether the point is disabled for use. The operationalState
reflects that of the MO referred to by the networkNode and is disabled if the
administrativeState of that MO is Locked";;
ATTRIBUTES
endPointId   GET,
networkNode   GET-REPLACE,
location   GET-REPLACE,
"ISO/IEC 10165-2":operationalState   GET,
"ISO/IEC 10165-2":administrativeState   GET-REPLACE;
NOTIFICATIONS
"ISO/IEC 10165-2":objectCreation,
"ISO/IEC 10165-2":objectDeletion,
"ISO/IEC 10165-2":attributeValueChange;;;
REGISTERED AS {vpnMO 3};

"Prep-Ext Vpn":vpnLink MANAGED OBJECT CLASS
DERIVED FROM "ISO/IEC 10165-2":top;
CHARACTERIZED BY
linkPkg PACKAGE
BEHAVIOUR
linkBeh BEHAVIOUR
DEFINED AS
"A link represents a generic communications link. The link defines the topology of
the communications link which may be unidirectional point-to-point, bidirectional
point-to-point, unidirectional point- to-multipoint, or multipoint-to-multipoint.
srcEndPoints indicate the sources of communications traffic across the link. These
sources may be only single end points. dstEndPoint indicates the sinks of
communications traffic across the link. These sink may be single end points or groups

of end points. Each sink point or group of sink points will receive traffic from every
source point. The qos attribute is a reference to either a qosSpec MO or a qosProfile
MO. The former contains a single QoS specification that should be applied to the
link. The latter contains references to several different qosSpec MOs together with
details of how the different QoS specifications they contain should be applied over
time. The link cannot be used if the administrativeState is Locked. The
operationalState indicates whether the link is disabled for use.";;
ATTRIBUTES
linkId   GET,
srcEndPoints   GET-REPLACE ADD-REMOVE,
dstEndPoints   GET-REPLACE ADD-REMOVE,
qos GET-REPLACE,
"ISO/IEC 10165-2":operationalState   GET,
"ISO/IEC 10165-2":administrativeState   GET-REPLACE;
NOTIFICATIONS
"ISO/IEC 10165-2":objectCreation,
"ISO/IEC 10165-2":objectDeletion,
"ISO/IEC 10165-2":stateChange,
"ISO/IEC 10165-2":attributeValueChange;;;
REGISTERED AS {vpnMO 4};

"Prep-Ext Vpn":networkLink MANAGED OBJECT CLASS
DERIVED FROM "Prep-Ext Vpn":vpnLink;
CHARACTERIZED BY
networkLinkPkg PACKAGE
BEHAVIOUR
networkLinkBeh BEHAVIOUR
DEFINED AS
"The networkLink is a link that represents resource reservation over a network domain
or set of network domains of interest. Its end points can be either the end of a chain of
network links or a translation end point connecting this network link to a network
link in a neighbouring domain.";;
ATTRIBUTES
       customerPathRelation   GET-REPLACE;;;
REGISTERED AS {vpnMO 5};

"Prep-Ext Vpn":networkLinkRelation MANAGED OBJECT CLASS
DERIVED FROM "ISO/IEC 10165-2":top;
CHARACTERIZED BY
networkLinkRelationPkg PACKAGE
BEHAVIOUR
networkLinkRelationBeh BEHAVIOUR
DEFINED AS
"networkLinkRelation relates two network links. One in the local network domain and
one in an adjacent network domain. It indicates through specialisation the maximum
QoS that can be allocated over the networkLink via this point. A specialisation
provides for different expressions of QoS suited to specific technologies and/or
policies";;
ATTRIBUTES
networkLinkRelationId   GET,

localNetworkLink   GET-REPLACE,
remoteNetworkLink   GET-REPLACE;
NOTIFICATIONS
"ISO/IEC 10165-2":objectCreation,
"ISO/IEC 10165-2":objectDeletion,
"ISO/IEC 10165-2":stateChange,
"ISO/IEC 10165-2":attributeValueChange;;;
REGISTERED AS {vpnMO 6};

"Prep-Ext Vpn":qosProfile MANAGED OBJECT CLASS
DERIVED FROM "ISO/IEC 10165-2":top;
CHARACTERIZED BY
qosProfilePkg PACKAGE
BEHAVIOUR
qosProfileBeh BEHAVIOUR
 DEFINED AS
"The qosProfile allows the QoS applied to a link to be specified in such a way that
different QoS specifications can be applied at different periods of time. The
periodicProfileList attribute contains a list of QoS specifications and the time from
which they should be in effect within a profile cycle. The length of the profile cycle is
given in the profileCycle attribute. The attribute oneOffProfileList is a list of QoS
specifications and the times between which they should be in effect. Between these
time the QoS specification is the one applied to the link instead of that specified by
the periodicProfileList";;
ATTRIBUTES
qosProfileId   GET,
profileCycle   GET-REPLACE,
periodicProfileList   GET-REPLACE ADD-REMOVE,
oneOffProfileList   GET-REPLACE ADD-REMOVE;;;
REGISTERED AS {vpnMO 7};

"Prep-Ext Vpn":qosSpec MANAGED OBJECT CLASS
DERIVED FROM "ISO/IEC 10165-2":top;
CHARACTERIZED BY
qosSpecPkg PACKAGE
BEHAVIOUR
 qosSpecBeh BEHAVIOUR
DEFINED AS
"The qosSpec contains the QoS parameters to be applied to a link. Technology-
specific QoS parameters are contained in MOs derived from this one";;
ATTRIBUTES
qosSpecId   GET;
NOTIFICATIONS
"ISO/IEC 10165-2":objectCreation,
"ISO/IEC 10165-2":objectDeletion,
"ISO/IEC 10165-2":stateChange,
"ISO/IEC 10165-2":attributeValueChange;;;
REGISTERED AS {vpnMO 8};

"Prep-Ext Vpn":terminationPoint MANAGED OBJECT CLASS
DERIVED FROM "Prep-Ext Vpn":endPoint;
CHARACTERIZED BY
endPointPkg PACKAGE
BEHAVIOUR
endPointBeh BEHAVIOUR
DEFINED AS
"A terminationPoint is an endPoint that represents the termination of a link at a point
where traffic inputs to and outputs from a network application";;;;
REGISTERED AS {vpnMO 9};

"Prep-Ext Vpn":translationPoint MANAGED OBJECT CLASS
DERIVED FROM "Prep-Ext Vpn":endPoint;
CHARACTERIZED BY
translationPointPkg PACKAGE
BEHAVIOUR
translationPointBeh BEHAVIOUR
DEFINED AS
"A translation point is an endPoint that additionally represents the translations from
network link in the local network domain to those in an adjacent one. ";;;;
REGISTERED AS {vpnMO 10};

"Prep-Ext Vpn":userStream MANAGED OBJECT CLASS
DERIVED FROM "Prep-Ext Vpn":vpnLink;
CHARACTERIZED BY
userStreamPkg PACKAGE
BEHAVIOUR
userStreamBeh BEHAVIOUR
DEFINED AS
"The userStream is a link that represents the end-to-end resource allocation
requirements of users in the VPN customer's domain. It defines a group of end points
between which communication has been enabled at a specific QoS.";;;;
REGISTERED AS {vpnMO 11};

E.2.1.2   Name Bindings

"Prep-Ext Vpn":customerPath-system NAME BINDING
SUBORDINATE OBJECT CLASS "Prep-Ext Vpn":customerPath;
NAMED BY SUPERIOR OBJECT CLASS "ISO/IEC 10165-2":system;
WITH ATTRIBUTE linkId;
CREATE WITH-REFERENCE-OBJECT;
DELETE ONLY-IF-NO-CONTAINED-OBJECTS;
REGISTERED AS {vpnNB 1};

"Prep-Ext Vpn":networkLink-system NAME BINDING
SUBORDINATE OBJECT CLASS "Prep-Ext Vpn":networkLink;
NAMED BY SUPERIOR OBJECT CLASS "ISO/IEC 10165-2":system;
WITH ATTRIBUTE linkId;
CREATE WITH-REFERENCE-OBJECT;
DELETE ONLY-IF-NO-CONTAINED-OBJECTS;

REGISTERED AS {vpnNB 10};

"Prep-Ext Vpn":qosProfile-system NAME BINDING
SUBORDINATE OBJECT CLASS "Prep-Ext Vpn":qosProfile;
NAMED BY SUPERIOR OBJECT CLASS "ISO/IEC 10165-2":system;
WITH ATTRIBUTE qosProfileId;
CREATE WITH-REFERENCE-OBJECT;
DELETE ONLY-IF-NO-CONTAINED-OBJECTS;
REGISTERED AS {vpnNB 12};

"Prep-Ext Vpn":terminationPoint-system NAME BINDING
SUBORDINATE OBJECT CLASS "Prep-Ext Vpn":terminationPoint;
NAMED BY SUPERIOR OBJECT CLASS "ISO/IEC 10165-2":system;
WITH ATTRIBUTE endPointId;
CREATE WITH-REFERENCE-OBJECT;
DELETE ONLY-IF-NO-CONTAINED-OBJECTS;
REGISTERED AS {vpnNB 4};

"Prep-Ext Vpn":translationPoint-system NAME BINDING
SUBORDINATE OBJECT CLASS "Prep-Ext Vpn":translationPoint;
NAMED BY SUPERIOR OBJECT CLASS "ISO/IEC 10165-2":system;
WITH ATTRIBUTE endPointId;
CREATE WITH-REFERENCE-OBJECT;
DELETE ONLY-IF-NO-CONTAINED-OBJECTS;
REGISTERED AS {vpnNB 6};

"Prep-Ext Vpn":userStream-system NAME BINDING
SUBORDINATE OBJECT CLASS "Prep-Ext Vpn":userStream;
NAMED BY SUPERIOR OBJECT CLASS "ISO/IEC 10165-2":system;
WITH ATTRIBUTE linkId;
CREATE WITH-REFERENCE-OBJECT;
DELETE ONLY-IF-NO-CONTAINED-OBJECTS;
REGISTERED AS {vpnNB 8};

E.2.1.3   Attributes

"Prep-Ext Vpn":connectionDescriptorId ATTRIBUTE
WITH ATTRIBUTE SYNTAX Prep-Ext-ASN1Module.SimpleNameType;
MATCHES FOR EQUALITY;
REGISTERED AS {vpnATT 1};

"Prep-Ext Vpn":customerPathRelation ATTRIBUTE
WITH ATTRIBUTE SYNTAX Prep-Ext-ASN1Module.DistinguishedName;
MATCHES FOR EQUALITY;
BEHAVIOUR
customerPathRelationBeh BEHAVIOUR
DEFINED AS
"This is a reference to the customerPath that the networkLink was created to
support.";;
REGISTERED AS {vpnATT 4};

"Prep-Ext Vpn":dstEndPoints ATTRIBUTE
WITH ATTRIBUTE SYNTAX Prep-Ext-ASN1Module.DestinationEndPoints;
MATCHES FOR EQUALITY;
BEHAVIOUR
dstEndPointsBeh BEHAVIOUR
DEFINED AS
"The list of end points that acts as the source of traffic in describing a link's topology.
The list elements may be either single end points or groups of end points.";;
REGISTERED AS {vpnATT 5};

"Prep-Ext Vpn":endPointId ATTRIBUTE
WITH ATTRIBUTE SYNTAX Prep-Ext-ASN1Module.SimpleNameType;
MATCHES FOR EQUALITY;
REGISTERED AS {vpnATT 6};

"Prep-Ext Vpn":linkId ATTRIBUTE
WITH ATTRIBUTE SYNTAX Prep-Ext-ASN1Module.SimpleNameType;
MATCHES FOR EQUALITY;
REGISTERED AS {vpnATT 7};

"Prep-Ext Vpn":localEndPointList ATTRIBUTE
WITH ATTRIBUTE SYNTAX Prep-Ext-ASN1Module.EndPointList;
MATCHES FOR EQUALITY;
BEHAVIOUR
localEndPointListBeh BEHAVIOUR
DEFINED AS
"localEndPointList are the end points in the local network domain whose connection
the connectionDescriptor is supporting";;
REGISTERED AS {vpnATT 8};

"Prep-Ext Vpn":location ATTRIBUTE
WITH ATTRIBUTE SYNTAX Prep-Ext-ASN1Module.DistinguishedName;
MATCHES FOR EQUALITY;
BEHAVIOUR
locationBeh BEHAVIOUR
DEFINED AS
"This is a technology independent address for the endPoint, implemented as a
distinguished name";;
REGISTERED AS {vpnATT 28};

"Prep-Ext Vpn":networkLinkUsed ATTRIBUTE
WITH ATTRIBUTE SYNTAX Prep-Ext-ASN1Module.ObjectInstance;
MATCHES FOR EQUALITY;
BEHAVIOUR
networkLinkUsedBeh BEHAVIOUR
DEFINED AS
"networkLink from which the resources for the userStream that the
connectionDescriptor is supporting should be allocated";;
REGISTERED AS {vpnATT 10};

"Prep-Ext Vpn":localNetworkLink ATTRIBUTE
WITH ATTRIBUTE SYNTAX Prep-Ext-ASN1Module.ObjectInstance;
MATCHES FOR EQUALITY;
BEHAVIOUR
localNetworkLinkBeh BEHAVIOUR
DEFINED AS
"The related networkLink which is in the same domain as the endPoint containing the
netWorkLinkRelationship";;
REGISTERED AS {vpnATT 11};

"Prep-Ext Vpn":networkLinkRelationId ATTRIBUTE
WITH ATTRIBUTE SYNTAX Prep-Ext-ASN1Module.SimpleNameType;
MATCHES FOR EQUALITY;
REGISTERED AS {vpnATT 12};

"Prep-Ext Vpn":networkNode ATTRIBUTE
WITH ATTRIBUTE SYNTAX Prep-Ext-ASN1Module.DistinguishedName;
MATCHES FOR EQUALITY;
BEHAVIOUR
networkNodeBeh BEHAVIOUR
DEFINED AS
"A networkNode provides the distinguished name of an object that represents the
physical network node onto which the end point is mapped.";;
REGISTERED AS {vpnATT 13};

"Prep-Ext Vpn":oneOffProfileList ATTRIBUTE
WITH ATTRIBUTE SYNTAX Prep-Ext-ASN1Module.OneOffProfileList;
MATCHES FOR EQUALITY;
BEHAVIOUR
oneOffProfileListBeh BEHAVIOUR
DEFINED AS
"A list of references to qosSpec MOs together with corresponding start and stop times
between which the parameters contained in the qosSpec MO should be used to
determine the profile's applied QoS.";;
REGISTERED AS {vpnATT 14};

"Prep-Ext Vpn":periodicProfileList ATTRIBUTE
WITH ATTRIBUTE SYNTAX Prep-Ext-ASN1Module.PeriodicProfileList;
MATCHES FOR EQUALITY;
BEHAVIOUR
periodicProfileListBeh BEHAVIOUR
DEFINED AS
"A list of references to a qosSpec MO and corresponding time values, taken from the
beginning of the cycle, from when the parameter contained in a qosSpec MO should
be used to determine the profile's applied QoS.";;
REGISTERED AS {vpnATT 15};

"Prep-Ext Vpn":profileCycle ATTRIBUTE
WITH ATTRIBUTE SYNTAX Prep-Ext-ASN1Module.ProfileCycle;

MATCHES FOR EQUALITY;
BEHAVIOUR
profileCycleBeh BEHAVIOUR
DEFINED AS
"Defines the period over which a pattern of QoS parameters is repeated.";;
REGISTERED AS {vpnATT 16};

"Prep-Ext Vpn":qos ATTRIBUTE
WITH ATTRIBUTE SYNTAX Prep-Ext-ASN1Module.DistinguishedName;
MATCHES FOR EQUALITY;
BEHAVIOUR
qosBeh BEHAVIOUR
DEFINED AS
"This attribute provides a pointer to an MO containing the specific QoS profile
information for the link.";;
REGISTERED AS {vpnATT 17};

"Prep-Ext Vpn":qosProfileId ATTRIBUTE
WITH ATTRIBUTE SYNTAX Prep-Ext-ASN1Module.SimpleNameType;
MATCHES FOR EQUALITY;
REGISTERED AS {vpnATT 18};

"Prep-Ext Vpn":qosSpecId ATTRIBUTE
WITH ATTRIBUTE SYNTAX Prep-Ext-ASN1Module.SimpleNameType;
MATCHES FOR EQUALITY;
REGISTERED AS {vpnATT 19};

"Prep-Ext Vpn":remoteEndPointList ATTRIBUTE
WITH ATTRIBUTE SYNTAX Prep-Ext-ASN1Module.EndPointList;
MATCHES FOR EQUALITY;
BEHAVIOUR
remoteEndPointListBeh BEHAVIOUR
DEFINED AS
"remoteEndPointList are the end points in a remote domain whose connection the
connectionDescriptor is supporting";;
REGISTERED AS {vpnATT 20};

"Prep-Ext Vpn":remoteNetworkLink ATTRIBUTE
WITH ATTRIBUTE SYNTAX Prep-Ext-ASN1Module.DistinguishedName;
MATCHES FOR EQUALITY;
BEHAVIOUR
remoteNetworkLinkBeh BEHAVIOUR
DEFINED AS
"The related network link in a domain other than the one the endPoint containing the
networkRelationship is in.";;
REGISTERED AS {vpnATT 21};

"Prep-Ext Vpn":srcEndPoints ATTRIBUTE
WITH ATTRIBUTE SYNTAX Prep-Ext-ASN1Module.EndPointList;
MATCHES FOR EQUALITY;

BEHAVIOUR
srcEndPointsBeh BEHAVIOUR
DEFINED AS
"The list of end points that acts as the source of traffic in describing a link's
topology";;
REGISTERED AS {vpnATT 22};

"Prep-Ext Vpn":userStreamSupported ATTRIBUTE
WITH ATTRIBUTE SYNTAX Prep-Ext-ASN1Module.ObjectInstance;
MATCHES FOR EQUALITY;
BEHAVIOUR
userStreamSupportedBeh BEHAVIOUR
DEFINED AS
"userStreamSupported is the user stream for which the connectionDescriptor
information is provided";;
REGISTERED AS {vpnATT 27};

E.2.2 ATM CBR Service Specific VPN Management Information Model

E.2.2.1 MO Schema Definitions

"Prep-Ext AtmCbrVpn":atmCbrSpec MANAGED OBJECT CLASS
DERIVED FROM "Prep-Ext Vpn":qosSpec;
CHARACTERIZED BY
atmCbrSpecPkg PACKAGE
BEHAVIOUR
atmCbrSpecBeh BEHAVIOUR
DEFINED AS
"The atmCbrSpec give the QoS parameters for a CBR service. These are maximum
bandwidth and cell delay variation";;
ATTRIBUTES
maximumBandwidth   GET-REPLACE,
cellDelayVariance   GET-REPLACE;;;
REGISTERED AS {atmCbrVpnMO 1};

"Prep-Ext AtmCbrVpn":cbrNetworkLinkRelation MANAGED OBJECT CLASS
DERIVED FROM "Prep-Ext Vpn":networkLinkRelation;
CHARACTERIZED BY
cbrNetworkLinkRelationPkg PACKAGE
BEHAVIOUR
cbrNetworkLinkRelationBeh BEHAVIOUR
DEFINED AS
"The cbrNetworkLinkRelation is a specialisation of networkLinkRelation for ATM
CBR networks. The maximum QoS that can be allocated to the local networkLink is
given in terms of a maximum bandwidth";;
ATTRIBUTES
maximumBandwidth   GET-REPLACE;;;
REGISTERED AS {atmCbrVpnMO 2};

"Prep-Ext AtmCbrVpn":pvcInterface MANAGED OBJECT CLASS
DERIVED FROM "Prep-Ext Vpn":connectionDescriptor;
CHARACTERIZED BY
pvcInterfacePkg PACKAGE
BEHAVIOUR
pvcInterfaceBeh BEHAVIOUR
DEFINED AS
"The pvcInterface is pointed to by the ConnectionDetails part of a
ConnectionDescriptor found in the connectionList attribute of an end point. It gives
the virtual chennel identifier and virtual path identifier values to be used at the link
represented by the end point in connecting to the userStream in the
ConnectionDescriptor to the remote endPoint in the ConnectionDescriptor.";;
ATTRIBUTES
vci   GET-REPLACE,
vpi   GET-REPLACE,
pvcDirection   GET-REPLACE;;;
REGISTERED AS {atmCbrVpnMO 3};

E.2.2.2   Name Bindings

"Prep-Ext AtmCbrVpn":atmCbrSpec-system NAME BINDING
SUBORDINATE OBJECT CLASS "Prep-Ext Vpn":atmCbrSpec;
NAMED BY SUPERIOR OBJECT CLASS "ISO/IEC 10165-2":system;
WITH ATTRIBUTE qosSpecId;
CREATE WITH-REFERENCE-OBJECT;
DELETE ONLY-IF-NO-CONTAINED-OBJECTS;
REGISTERED AS {atmCbrVpnNB 1};

"Prep-Ext AtmCbrVpn":cbrNetworkLinkRelation-translationPoint NAME BINDING
SUBORDINATE OBJECT CLASS "Prep-Ext atmCbrVpn":cbrNetworkLinkRelation;
NAMED BY SUPERIOR OBJECT CLASS "Prep-Ext Vpn":translationPoint;
WITH ATTRIBUTE networkLinkRelationId;
CREATE WITH-REFERENCE-OBJECT;
DELETE ONLY-IF-NO-CONTAINED-OBJECTS;
REGISTERED AS {atmCbrVpnNB 3};

"Prep-Ext AtmCbrVpn":pvcInterface-terminationPoint NAME BINDING
SUBORDINATE OBJECT CLASS "Prep-Ext atmCbrVpn":pvcInterface;
NAMED BY SUPERIOR OBJECT CLASS "Prep-Ext Vpn":terminationPoint;
WITH ATTRIBUTE connectionDescriptorId;
CREATE WITH-REFERENCE-OBJECT;
DELETE ONLY-IF-NO-CONTAINED-OBJECTS;
REGISTERED AS {atmCbrVpnNB 4};

"Prep-Ext AtmCbrVpn":pvcInterface-translationPoint NAME BINDING
SUBORDINATE OBJECT CLASS "Prep-Ext atmCbrVpn":pvcInterface;
NAMED BY SUPERIOR OBJECT CLASS "Prep-Ext Vpn":translationPoint;
WITH ATTRIBUTE connectionDescriptorId;
CREATE WITH-REFERENCE-OBJECT;

DELETE ONLY-IF-NO-CONTAINED-OBJECTS;
REGISTERED AS {atmCbrVpnNB 5};

E.2.2.3   Attributes

"Prep-Ext AtmCbrVpn":cellDelayVariance ATTRIBUTE
WITH ATTRIBUTE SYNTAX Prep-Ext-ASN1Module.CellDelayVariance;
MATCHES FOR EQUALITY;
BEHAVIOUR
cellDelayVarianceBeh BEHAVIOUR
DEFINED AS
"measurement of cell delay variance";;
REGISTERED AS {atmCbrVpnATT 2};

"Prep-Ext AtmCbrVpn":maximumBandwidth ATTRIBUTE
WITH ATTRIBUTE SYNTAX Prep-Ext-ASN1Module.MaximumBandwidth;
MATCHES FOR EQUALITY;
BEHAVIOUR
maximumBandwidthBeh BEHAVIOUR
DEFINED AS
"measurement of maximum bandwidth";;
REGISTERED AS {atmCbrVpnATT 3};

"Prep-Ext AtmCbrVpn":pvcDirection ATTRIBUTE
WITH ATTRIBUTE SYNTAX Prep-Ext-ASN1Module.PvcDirection;
MATCHES FOR EQUALITY;
BEHAVIOUR
pvcDirectionBeh BEHAVIOUR
DEFINED AS
"Defines whether the traffic along the pvc connecting to the pvcInterface is passing
traffic into the local network, out of the local network, or both.";;
REGISTERED AS {atmCbrVpnATT 4};

"Prep-Ext AtmCbrVpn":vci ATTRIBUTE
WITH ATTRIBUTE SYNTAX Prep-Ext-ASN1Module.Vci;
MATCHES FOR EQUALITY;
BEHAVIOUR
vciBeh BEHAVIOUR
DEFINED AS
"Defines the virtual channel identifier value for a virtual circuit";;
REGISTERED AS {atmCbrVpnATT 6};

"Prep-Ext AtmCbrVpn":vpi ATTRIBUTE
WITH ATTRIBUTE SYNTAX Prep-Ext-ASN1Module.Vpi;
MATCHES FOR EQUALITY;
BEHAVIOUR
vpiBeh BEHAVIOUR
DEFINED AS
"Defines the virtual path identifier value for a virtual circuit";;
REGISTERED AS {atmCbrVpnATT 7};

E.2.2.4   ASN.1

EndPointList ::= SET OF DistinguishedName

DestinationEndPoints ::= CHOICE {
        endPointList   [0] EndPointList,
        endPointListSet   [1] SET OF EndPointList }

ConnectionList ::= SET OF ObjectInstance

TranslationList ::= SET OF ObjectInstance

ProfileCycle ::= ENUMERATED {
        calenderYear   (0),
        calenderMonth   (1),
        week   (2),
        day   (3),
        hour   (4) }

PeriodicProfileList ::= SET OF PeriodicProfile

PeriodicProfile ::= SEQUENCE {
startTime   StartTime,
qosSpecification   ObjectInstance }

OneOffProfileList ::= SET OF OneOffProfile

OneOffProfile ::= SEQUENCE {
        startTime   StartTime,
        stopTime   StopTime,
        qosSpecification   ObjectInstance }

Vpi ::= INTEGER

Vci ::= INTEGER

PvcDirection ::= ENUMERATED {
        in   (0),
        out   (1),
        inAndOut   (2) }

MaximumBandwidth ::= INTEGER

CellDelayVariance ::= INTEGER

PREPARE Papers on Inter-Domain Management

This section provides details of the papers published by PREPARE on its work in the area of inter-domain management.

Bjerring, L.H., "Flexible management of end-to-end services," in *Proceedings RACE International Conference on Intelligence in Broadband Services and Networks, Paris, November 1993*, pp. II/5/1-10.

Bjerring, L.H. and J.M. Schneider, "End-to-end Service Management with Multiple Providers," in *Towards a Pan-European Telecommunication Service Infrastructure - IS&N '94. Proceedings Second International Conference on Intelligence in Broadband Services and Networks, Aachen, Germany, September 7-9, 1994*, ed. H.-J. Kugler, A. Mullery, and N. Niebert, pp. 305-318, Springer-Verlag, Berlin, 1994.

Bjerring, L.H. and M. Tschichholz, "Requirements of Inter-Domain Management and their Implications for TMN Architecture and Implementation," in *Towards a Pan-European Telecommunication Service Infrastructure - IS&N '94. Proceedings Second International Conference on Intelligence in Broadband Services and Networks, Aachen, Germany, September 7-9 1994*, ed. H.-J. Kugler, A. Mullery, and N. Niebert, pp. 193-206, Springer-Verlag, Berlin, 1994.

Hall, J., I. Schieferdecker, and M. Tschichholz, "Customer requirements on teleservice management," in *Integrated Network Management IV, Proceedings of the Fourth International Symposium on Integrated Network Management, 1995*, ed. A.S. Sethi, Y. Raynaud, and F. Faure-Vincent, pp. 143-155, Chapman & Hall, London, 1995.

Kisker, W. and J.M. Schneider, "ATM Public Network Management in PREPARE", in *Bringing Telecommunication Services to the People - IS&N'95. Third International Conference on Intelligence in Broadband Services and Networks. Heraklion, Crete, Greece, October 1995, Proceedings*, ed. A. Clarke, M. Campolargo, and N. Karatzas, pp. 261-274, Springer Verlag, Berlin, 1995.

Lewis, D. and J. Crowcroft, "Multimedia Applications and Services in the PREPARE Testbed," in *Proceedings 2nd International Conference on Broadband Islands, Athens, June 1993*, pp. 235-238.

Lewis, D., "Layered management services provide support for multimedia conferencing," in *Proceedings RACE International Conference on Intelligence in Broadband Services and Networks, Paris, November 1993*, pp. IX/1/1-10.

Lewis, D., L.H. Bjerring, and I.H. Thorarensen, "An Inter-domain Virtual Private Network Management Service," in *Proceedings of the IEEE Network Operations and Management Symposium. Kyoto, Japan, April 15-19, 1996*, (NOMS'96), pp. 115-123, IEEE Communications Society, Piscataway, NJ, 1996.

Lewis, D., W. Donnelly, J.M. Schneider, and M. Klotz, "Managing Broadband VPN Services in the PREPARE Testbed," in *Proceedings RACE International Conference on Intelligence in Broadband Services and Networks, Paris, November 1993*, pp. II/3/1-15.

Lewis, D. and P. Kirstein, "A Broadband Testbed for the Investigation of Multimedia Services and Teleservice Management," in *Proceedings 3rd International Conference on Broadband Islands, Hamburg, April 1994*.

Lewis, D., S. O'Connell, W. Donnelly, and L. Bjerring, "Experiences in multi-domain management system development," in *Integrated Network Management IV, Proceedings of the Fourth International Symposium on Integrated Network Management, 1995*, ed. A.S. Sethi, Y. Raynaud, and F. Faure-Vincent, pp. 494-505, Chapman & Hall, London, 1995.

Lewis, D., T. Tiropanis, L. Bjerring, and J. Hall, "Experiences in Multi-domain Management Service Development", in *Bringing Telecommunication Services to the People - IS&N'95. Third International Conference on Intelligence in Broadband Services and Networks. Heraklion, Crete, Greece, October 1995, Proceedings*, ed. A. Clarke, M. Campolargo, and N. Karatzas, pp. 174-184, Springer Verlag, Berlin, 1995.

Nielsen, B.H., "PREPARE - Management of Heterogeneous Broadband Testbed," *Teleteknik* (English Edition), 1994, No. 2, pp. 77-87.

Nielsen, P.S. and B. Lonvig, "Development of Telecommunications Management Systems using OO Methods and CASE Tool Support," in *Towards a Pan-European Telecommunication Service Infrastructure - IS&N '94. Proceedings Second International Conference on Intelligence in Broadband Services and Networks, Aachen, Germany, September 7-9, 1994*, ed. H.-J. Kugler, A. Mullery, and N. Niebert, pp. 407-418, Springer-Verlag, Berlin, 1994.

O'Connell, S. and W. Donnelly, "Security Requirements of the TMN X-Interface within End-to-End Service Management of Virtual Private Networks," in *Towards a Pan-European Telecommunication Service Infrastructure - IS&N '94. Proceedings Second International Conference on Intelligence in Broadband Services and Networks, Aachen, Germany, September 7-9 1994*, ed. H.-J. Kugler, A. Mullery, and N. Niebert, pp. 207-217, Springer-Verlag, Berlin, 1994.

Schneider, J.M. and W. Donnelly, "An Open Architecture for Inter-Domain Communications Management in the PREPARE Testbed," in *Proceedings 2nd International Conference on Broadband Islands, Athens, June 1993*, pp. 77-88.

Schneider, J.M. and W. Donnelly, "Cooperative Management of Integrated Broadband Communication Services over the CCITT "X" Interface," in *Proceedings 3rd International Workshop on Distributed Systems, Operations and Management (DSOM), Long Branch, NJ, October 1993*.

Schneider, J.M., T. Preuß, and P.S. Nielsen, "Management of Virtual Private Network Services for Integrated Broadband Communication", *Computer Communication Review*, vol. 23 no. 4, pp. 224-237, October 1993.

Schneider, J.M. and W. Donnelly, "End-to-end Communications Management using TMN "X" Interfaces," *Journal of Network and Systems Management*, vol. 3, no. 1, pp. 85-110, March 1995.

Tschichholz, M. and W. Donnelly, "The PREPARE Management Information Service," in *Proceedings RACE International Conference on Intelligence in Broadband Services and Networks, Paris, November 1993* , pp. IV/3/1-12.

Tschichholz, M., J. Hall, S. Abeck, and R. Wies "Information Aspects and Future Directions in an Integrated Telecommunications and Enterprise Management Environment," *Journal of Network and Systems Management*, vol. 3, no. 1, pp. 111-138, March 1995.

References

[90/387/EEC]   *Council Directive of 28 June 1990 on the establishment of the internal market for telecommunications services through the implementation of open network provision (90/387/EEC).*

[Albanese]   Albanese, A., et. al., "An Enterprise Network Management System Architecture," in *Proceedings of the Fifth RACE TMN Conference, London, November 1991*, pp. D3Plen/4/1-10, 1991.

[Alpers]   Alpers, B., et. al., "DOMAINS - Basic Concepts for Management of Distributed Systems," in *Proceedings of the Fifth RACE TMN Conference, London, November 1991*, pp. III.2/5/1-15, 1991.

[Bangemann]   *Europe and the global information society. Recommendations to the European Council* (Bangemann Report), European Commission, Brussels, May 1994.

[Basu]   Basu, K.,"Open Network Architecture and Information Services: Teletraffic Challenges," *Computer Networks and ISDN Systems*, vol. 20, nos. 1-5, pp. 101-107, December 1990.

[BenPol]   Bennett, R.L. and G.E. Policello II, "Switching Systems in the 21st Century," *IEEE Communications Magazine*, vol. 31, no. 6, pp. 24-28, March 1993.

[Bernstein]   Bernstein, P.A., Hadzilacos, V., Goodman, N., *Concurrency Control and Recovery in Database Systems*, Addison-Wesley Publishing Company, Reading, Ma., 1987.

[BernYuh]   Bernstein, L. and C.M. Yuhas, "Can we talk?" in *Integrated Network Management IV, Proceedings of the fourth international symposium on integrated network management, 1995*, ed. A.S. Sethi, Y. Raynaud, and F. Faure-Vincent, pp. 670-676, Chapman & Hall, London, 1995.

[CBDS]   *CBDS, Connectionless Broadband Data Service.* ETSI Draft ETS, Basel version July 1991, (T/NA(91)21).

[CFS-H430]   *The Inter-Domain Management Information Service*, RACE Common Functional Specification H430, RACE Industrial Consortium, Brussels, 1994.

[Cooper]   Cooper, M., "Telecoms Technology to Delight Tomorrow's Customers," in *Speakers' Papers, 7th World Telecommunication Forum Technology Summit, 3-11 October 1995, vol. 1, "Convergence of Technologies, Services and Applications"*, pp. 379-383, ITU, Geneva, 1995.

[CORBA]   *The Common Object Request Broker: Architecture and Specification.* OMG Document Number 92.12.1, Revision 1.1. Object Management Group, 1992. **And:** *Object Request Broker 2.0.* Object Management Group, 1995.

[COSS]   *Common Object Services Specification*, Volumes 1 and 2, Object Management Group, 1995.

[DeBony]   De Bony, E., "Global Networks and Interoperability," *I&T Magazine*, European Commission, pp. 11-13, Winter 1994-95.

[Ejiri]   Ejiri, M., "The paradigm shift in telecommunications services and networks," in *Integrated Network Management IV, Proceedings of the fourth international symposium on integrated network management, 1995*, ed. A.S. Sethi, Y. Raynaud, and F. Faure-Vincent, pp. 688-699, Chapman & Hall, London, 1995.

[EliaVeij]   Eliassen, F. and J. Veijalainen, *An S-Transaction Definition Language and Execution Mechanism*, Arbeitspapiere der GMD 275, GMD, Sankt Augustin, 1987.

[ETR230]   *Network Aspects - Telecommunications Management Network (TMN) - TMN Standardisation Overview*, ETSI Draft ETR230, Version 1.0, 1995.

[G.803]   *Architectures of transport networks based on the synchronous digital hierarchy (SDH)*, ITU-T Recommendation G.803, 1993.

[Gaarder]   Gaarder, J., *Sophie's World*, trans. P. Møller, Phoenix House London, 1995.

[GII]   *The Global Information Infrastructure: Agenda for Cooperation*, U.S. Department of Commerce, Washington, DC, 1995.

[Handley]   Handley, M.J, P.T. Kirstein, and M.J. Sasse, "Multimedia integrated conferencing for European researchers (MICE): piloting activities and the conference management and multiplexing centre," *Computer Networks and ISDN Systems*, vol. 26, no. 3, pp. 275-290, November 1993.

[HPNRG]   *Report of the High Performance Networking Requirements Group*, European Commission, April 1994.

[I.362]   *B-ISDN ATM Adaption Layer (AAL) functional description*. ITU-T Recommendation I.362, 1992.

[IEEE829]   *IEEE Standard for Software Test Documentation*, IEEE Standard 829-1983.

[ISO-8571]   *Information processing systems - Open Systems Interconnection - File Transfer Access and Management*. ISO/IEC International Standard 8571, 1988.

[KiskSchn]   Kisker, W. and J.M. Schneider, "ATM Public Network Management in PREPARE", in *Bringing Telecommunication Services to the People - IS&N'95. Third International Conference on Intelligence in Broadband Services and Networks. Heraklion, Crete, Greece, October 1995, Proceedings*, ed. A. Clarke, M. Campolargo, and N. Karatzas, pp. 261-274, Springer Verlag, Berlin, 1995.

[Kitson]   Kitson, B., "CORBA and TINA: The Architectural Relationships," in *TINA 95, Integrating Telecommunications and Distributed Computing - from Concepts to Reality, Conference Proceedings, Melbourne, February 1995*, pp. 371-386, 1995.

[Kvols]   Kvols, K., "Combine - Interworking on Broadband Networks," *Teleteknik* (English Edition), vol. 38, no. 2, pp. 65-76, 1994.

[Lengdell]   Lengdell, M., H. Oshigri, and J. Pavon, "The TINA Network Resource Model," in *Proceedings Globecom '95*.

[Lloyd]   Lloyd, P., "Outsourcing: Mutual Benefit or Mutual Risk?" *Telecommunications*, vol. 27, no. 2, pp. 37-42, February 1993.

[M.3000]   *Overview of TMN recommendations*, ITU-T Recommendation M.3000, 1995.

[M.3010]   *Principles for a Telecommunications Management Network*, ITU-T Recommendation M.3010, 1992.

[M.3010/95]   *Principles for a Telecommunications Management Network*, ITU-T Draft Revised Recommendation M.3010, 1995.

[M.3020/92]   *TMN Interface Specification Methodology*, ITU-T Recommendation M.3020, 1992.

[M.3020/94]   *TMN Interface Specification Methodology*, ITU-T Draft Revised Recommendation M.3020, 1994.

[M.3100]   *Generic Network Information Model*, ITU-T Recommendation M.3100, 1992.

[M.3200]   *TMN Management Services: Overview*, ITU-T Recommendation M.3200, 1992.

[McCarthy]   McCarthy, K., G. Pavlou, S. Bhatti, and J.N. de Souza, "Exploiting the power of OSI Management for the control of SNMP-capable resources using generic application level gateways," in *Integrated Network Management IV, Proceedings of the fourth international symposium on integrated network management, 1995*, ed. A.S. Sethi, Y. Raynaud, and F. Faure-Vincent, pp. 688-699, Chapman & Hall, London, 1995.

[Moeller]   Moeller, E. et al., "The BERKOM multimedia mail teleservice," *Computer Communications*, vol. 18, no. 2, pp. 89-102, February 1995.

[NA43308]   *Baseline Document on the Integration of IN and TMN*, ETSI NA43308, 1992.

[NA43316]   *TMN Generic Managed Object Library for the Network Level View*, ETSI NA43316, 1995.

[NA52210]   *B-ISDN Management Architecture and Management Information Model for the ATM crossconnect*, ETSI NA52210, 1995.

[Nielsen]   Nielsen, P.M., "Batman - Danish ATM Cooperation," *Teleteknik*, vol. 44, no. 3, pp. 149-161, 1993 (in Danish).

[NMF-025]   *The "Ensemble" Concepts and Format, Forum 025,* Issue 1.0, Network Management Forum, Morristown, NJ, 1992.

[NMF-026]   *Translation of Internet MIBs to ISO/CCITT GDMO MIBs, Forum 026,* Issue 1.0, Network Management Forum, Morristown, NJ, 1993.

[NMF-028]   *ISO/CCITT to Internet Management Proxy, Forum 028,* Network Management Forum, Morristown, NJ, 1993.

[NMF-BPM]   *A Service Management Business Process Model,* Network Management Forum, Morristown, NJ, 1995.

[NMF-CMIS++]   *CMIS++: CMISE and ACSE C++ Application Programming Interface,* NMF xxx, Issue 1.0, Draft 7, Network Management Forum, Morristown, NJ, 1996.

[NMF-SF]   *The OMNIPoint Strategic Framework. A Service-Based Approach to the Management of Networks and Systems,* Network Management Forum, Morristown, NJ, 1993.

[NMF-TMN]   *TMN C++ Application Programming Interface,* NMF xxx, Issue 1.0, Draft 5, Network Management Forum, Morristown, NJ, 1996.

[NMF-TR107]   *ISO/CCITT and Internet Management: Coexistence and Interworking Strategy, Forum TR107,* Network Management Forum, Morristown, NJ, 1992.

[NMF-TR114]   *OMNIPoint Integration Architecture, Forum TR114,* Network Management Forum, Morristown, NJ, 1995.

[NMF-TR115]   *An OMNIPoint TMN Design Guide. Using ISP's to Implement the Management Communication Aspects of TMN Q3 and X Interfaces, Forum TR115,* Issue 1.0, Network Management Forum, Morristown, NJ, 1995.

[ODMA]   *Open Distributed Management Architecture, Committee Draft.* ISO/IEC JTC1/SC21 N9988, November 1995.

[Okazaki]   Okazaki, Y, Y. Hibino, T. Togawa, and N. Akiyama, "TMN-Based Network Operations and Management of Switching Systems," in *Proceedings of the IEEE Network Operations and Management Symposium. Kissimmee, Florida, February 14-17, 1994,* (NOMS'94), pp. 754-765, IEEE Communications Society, Piscataway, NJ, 1994.

[OMA]   *Object Management Architecture Guide,* ed. R.M. Soley, OMG Document Number 92.11.1, Revision 2.0 Second Edition. Object Management Group, Framingham, MA, 1992.

[Onions]   Onions, J., "ISODE: In Support of Migration," *Computer Networks and ISDN Systems,* vol. 17, nos. 4/5, pp. 362-366, October 1989.

[Q.1204]   *Intelligent Network Distributed Functional Plane Architecture,* ITU-T Recommendation Q.1204, 1993.

[Q.1205]   *Intelligent Network Physical Plane Architecture*, ITU-T Recommendation Q.1205, 1993.

[Pavlou]   Pavlou, G., T. Tinn, and A. Carr, "High-Level Access APIs in the OSIMIS TMN Platform: Harnessing and Hiding", in *Towards a Pan-European Telecommunication Service Infrastructure - IS&N '94. Proceedings Second International Conference on Intelligence in Broadband Services and Networks, Aachen, Germany, September 7-9 1994*, ed. H.-J. Kugler, A. Mullery, and N. Niebert, pp. 219-230, Springer-Verlag, Berlin, 1994.

[Q.751]   *Network Element Management Information Model for the Message Transfer Part*, ITU-T Recommendation Q.751, 1995.

[RACE]   *Research and technology development in advanced communications technologies in Europe. RACE 1995*, Directorate General XIII, European Commission, Brussels 1995.

[Richter]   Richter, A., *Transaktionen im Interdomain-Management - Konzept für die Integration von Transaction-Processing in den Interdomain Management Information Service (IDMIS)*, Diplomarbeit, Technical University of Berlin, March 1994.

[Seitz]   Seitz, N.B., S. Wolf, S. Voran, and R. Bloomfield, "User-Oriented Measures of Telecommunication Quality," *IEEE Communications Magazine*, vol. 32, no. 1, pp. 56–66, January 1994.

[Sloman]   Sloman, M., J. Magee, K. Twidle, and J. Kramer, "An Architecture for Managing Distributed Systems," in *Proceedings of the Fourth Workshop on Future Trends of Distributed Computing Systems, September 22-24 1993, Lisbon, Portugal*, IEEE Computer Society, Los Alamitos, 1993.

[SloTwi]   Sloman, M. and K. Twidle, "Domains: A Framework for Structuring Management Policy," in *Network and Distributed Systems Management*, ed. M. Sloman, pp. 433-453, Addison-Wesley, Wokingham, 1994.

[SMART-Ord]   *Ordering White Paper*, Issue 1.0. NMF SMART Ordering Team, 1995.

[SMART-P-IF] *P Specification for Customer-to-SP Trouble Management*, Working version as of 26 June 1995, NMF SMART, 1995.

[SNMP]   Case, J.D., M. Fedor, M.L. Schoffstall, and C. Davin, *A Simple Network Management Protocol (SNMP)*. RFC 1157, May 1990.

[Snelling]   Snelling, R.K., "Prometheus Unbound," in *Proceedings of the IEEE Network Operations and Management Symposium. Kissimmee, Florida, February 14-17, 1994*, (NOMS'94), pp. 371-421, IEEE Communications Society, Piscataway, NJ, 1994.

[SPIRIT]   *X/Open Consortium Specification. SPIRIT Platform Blueprint*, SPIRIT Issue 2.0, Network Management Forum, Morristown, NJ, 1994.

| [StrenDob] | Strens, R. and J. Dobson, "Responsibility Modelling as a Technique for Organisational Requirements Definition," *Intelligent Systems Engineering*, vol. 3, no. 1, pp. 20-26, 1994. |

[TINA-005]   Graubmann, P. and N. Mercouroff, *Engineering Modelling Concepts (DPE Architecture)*, TINA Baseline document TB_NS.005_2.0_94, December 1994.

[TINA-010]   de la Fuente, L.A. and T. Walles, *TINA-C Management Architecture*. TINA Baseline document TB_GN.010_2.0_94, December 94.

[TINA-012]   Berndt, H. and R. Minerva, *Service Architecture*, TINA Baseline document TB_MDC.012_2.0_94, March 1995.

[TINA-018]   Chapman, M. and S. Montesi, *Overall Concepts and Principles of TINA*, TINA Baseline document TB_MDC.018_1.0_94, December 1994.

[TINA-020]   Janson, R. and J. Sallros, *Domain types and basic Reference Points in TINA*, Draft TINA Baseline document EN_RIC.020_0.5_95, May 1995.

[TMFA]   *Telecommunications Management Functional Areas*, RACE Common Functional Specifications H400-H414, RACE Industrial Consortium, Brussels, 1994.

[Vecchi]   Vecchi, M.P., "Broadband Networks and Services: Architecture and Control", *IEEE Communications Magazine*, vol. 33, no. 8, pp. 24-32, August 1995.

[Vetter]   Vetter, R.J., "ATM Concepts, Architectures, and Protocols," *Communications of the ACM*, vol. 38, no. 2, pp. 30-38, February 1995.

[Willetts]   Willetts, K., "Service Management - The Drive for Re-engineering," in *Proceedings of the IEEE Network Operations and Management Symposium. Kissimmee, Florida, February 14-17, 1994*, (NOMS'94), pp. 24-35, IEEE Communications Society, Piscataway, NJ, 1994.

[X.200]   *Information Processing Systems - Open Systems Interconnection - Basic Reference Model*, ISO/IEC International Standard, ITU-T Recommendation X.200 / ISO/IEC 7498-1, 1984.

[X.208]   *Information technology - Open Systems Interconnection - Specification of Abstract Syntax Notation One (ASN.1)*, ITU-T Recommendation X.208 / ISO/IEC International Standard 8824, 1990.

[X.400]   *Message Handling Services: Message Handling System and Service Overview*, ITU-T Recommendation X.400 / ISO/IEC International Standard 10021-1, 1993.

[X.407]   *Message Handling Systems: Abstract Service Definition Conventions*, ITU-T Recommendation X.407 / ISO/IEC International Standard 10021-3, 1988.

[X.500]   *Information technology - Open Systems Interconnection - The Directory: Overview of Concepts, Models, and Services*, ITU-T Recommendation X.500 / ISO/IEC International Standard 9594-1, 1993.

[X.660]   *Information technology - Open Systems Interconnection - Procedures for the operation of OSI Registration Authorities - Part 1: General procedures*, ITU-T Recommendation X.660 / ISO/IEC International Standard 9834-1, 1990.

[X.700]   *Management Framework for Open Systems Interconnection (OSI) for CCITT Applications*, ITU-T Recommendation X.700 / *Information processing systems - Open Systems Interconnection - Basic Reference Model - Part 4: Management Framework*, ISO/IEC International Standard 7498-4, 1989.

[X.701]   *Information technology - Open Systems Interconnection - Systems Management Overview*, ITU-T Recommendation X.701 / ISO/IEC International Standard 10040, 1992.

[X.702]   *Information technology - Open Systems Interconnection - Application Context for Systems Management with Transaction Processing*, ITU-T Draft Recommendation X.702, ISO/IEC Draft International Standard 11587, 1994.

[X.710]   *Common Management Information Service Definition [for CCITT Applications]*, ITU-T Recommendation X.710 / ISO/IEC International Standard 9595, 1991.

[X.711]   *Information technology - Open Systems Interconnection - Common Management Information Protocol Specification*, ITU-T Recommendation X.711 / ISO/IEC International Standard 9596-1, 1991.

[X.712]   *Information technology - Open Systems Interconnection - Common Management Information Protocol: Protocol Implementation Conformance Statement (PICS) proforma*, ITU-T Recommendation X.712 / ISO/IEC International Standard 9596-2, 1992.

[X.720]   *Information technology - Open Systems Interconnection - Structure of Management Information - Part 1: Management Information Model*, ITU-T Recommendation X.720 / ISO/IEC International standard 10165-1, 1993.

[X.721]   *Information technology - Open Systems Interconnection - Structure of Management Information - Part 2: Definition of Management Information*, ITU-T Recommendation X.721 / ISO/IEC International Standard 10165-2, 1992.

[X.722]   *Information technology - Open Systems Interconnection - Structure of Management Information - Part 4: Guidelines for the Definition of Managed Objects*, ITU-T Recommendation X.722 / ISO/IEC International Standard 10165-4, 1992.

[X.723]   *Information technology - Open Systems Interconnection - Structure of Management Information - Part 5: Generic Management Information*, ITU-T Recommendation X.723 / ISO/IEC International Standard 10165-5, 1993.

[X.724]   *Information technology - Open Systems Interconnection - Structure of Management Information - Part 6: Requirements and Guidelines for Implementation Conformance Statement Proformas*, ITU-T Recommendation X.724 / ISO/IEC International Standard 10165-6, 1993.

[X.725]   *Information technology - Open Systems Interconnection - Structure of Management Information - Part 7: General Relationship Model*, ITU-T Recommendation X.725 / ISO/IEC International Standard 10165-7, 1995.

[X.742]   *Information technology - Open Systems Interconnection - Systems Management: Usage Metering Function*, ITU-T Draft Recommendation X.742 / ISO/IEC Draft International Standard 10164-10, 1993.

[X.750]   *Information technology - Open Systems Interconnection - Systems Management: Management Knowledge Management Function*, ITU-T Draft Recommendation X.750 / ISO/IEC Draft International Standard 10164-16, 1994.

[X.790]   *Application Services:Trouble Management Function for ITU-T Applications*, ITU-T Recommendation X.790, 1995.

[X.851]   *Information Technology - Open Systems Interconnection - Service Definition for the Commitment, Concurrency and Recovery service element*, ITU-T Recommendation X.851 / ISO/IEC International Standard 9804, 1993.

[X.852]   *Information Technology - Open Systems Interconnection - Commitment, Concurrency and Recovery service element: Protocol specification*, ITU-T Recommendation X.852 (1993) / ISO/IEC International Standard 9805-1, 1994.

[X.860]   *Open Systems Interconnection - Distributed Transaction Processing: Model*, ITU-T. Recommendation X.860 / ISO/IEC International Standard 10026-1, 1991.

[X.861]   *Open Systems Interconnection - Distributed Transaction Processing: Service Definition*, ITU-T Recommendation X.861 / ISO/IEC International Standard 10026-2, 1991.

[X.862]   *Open Systems Interconnection - Distributed Transaction Processing: Protocol Specification*, ITU-T Recommendation X.862 / ISO/IEC International Standard 10026-3, 1991.

[X.901]   *Information Technology - Open Distributed Processing - Reference Model - Part 1: Overview*, ITU-T Draft Recommendation X.901 / ISO/IEC Draft International Standard 10746-1, 1995.

[X.902]   *Information Technology - Open Distributed Processing - Reference Model - Part 2: Foundations*, ITU-T Recommendation X.902 / ISO International Standard 10746-2, 1995.

[X.903]   *Information Technology - Open Distributed Processing - Reference Model - Part 3: Architecture*. ITU-T Recommendation X.903 / ISO International Standard 10746-3, 1995.

[X.904]   *Information Technology - Open Distributed Processing - Reference Model - Part 4: Architectural semantics*. ITU-T Draft Recommendation X.904 / ISO Draft International Standard 10746-4, 1995.

[XoJIDM]   *Inter-Domain Management Specifications: Specification Translation*, X/Open Preliminary Specification, X/Open, Reading, Draft of April 17, 1995.

Glossary

This glossary is intended to help the reader. It defines words and expressions which are related to the core subjects addressed by this book. Entries are alphabetically sorted.

Bearer Service: A bearer service is a service offered directly by a network, and thus offered by a network operator. Bearer services provide the basis for other types of service, in particular value-added services which add value "on top" of bearer services.

Cooperative Management: A management strategy where different stakeholders cooperate in order to achieve a defined management task. Cooperative management can be organised on either a peer-to-peer or a hierarchical basis (e.g., between an ATM service provider and a VPN service provider).

Customer: Customer denotes in the context of PREPARE an organisational stakeholder which uses all telecommunications networks and services provided by other organisational actors. For phase 1 of PREPARE the customer was instantiated by car manufacturer companies while in the second phase it was instantiated by a worldwide shipping company.

Domain: A domain is associated with an organisation. The set of resources, both virtual and real, belonging to an organisation comprises a management administrative domain, which is defined as a "management domain where the managed objects in the domain are all under the responsibility of one and only one administrative authority" [X.701].

End-to-End: The concept of end-to-end means that "something" (e.g., a telecommunications service, telecommunications management) is considered as a whole, even though it may be constructed or provided using multiple constituents distributed over several domains. For instance, an end-to-end worldwide telecommunications service is offered as a single item to its customers, or the end-to-end telecommunications management of an environment results from the management of each constituent domain.

End User: The end users use/receive telecommunications services. End users are in the customer's domain.

Enterprise Model: According to Part 1 of the RM-ODP [X.901] an enterprise model is a specification of a system as seen from an enterprise viewpoint. It is concerned with the business activities of the system. It covers business policies, human user roles with respect to the system under consideration and its environment. In PREPARE, enterprise modelling has been used as an approach to capture and analyse requirements on the PREPARE inter-domain management solutions (cf. chapter 4). The resulting enterprise model considers the stakeholders and roles, their environment, and the services exchanged between them.

Hierarchical Cooperation: Management cooperation where a stakeholder provides management information and/or services to another one. For instance, an ATM service provider reports statistical traffic information to its customers.

IBC (Integrated Broadband Communications): A concept developed by the RACE programme which aims at supporting the telecommunications requirements of the information society in Europe at the beginning of the 21st century. The IBC environment supports broadband data, audio, and image communication. It is characterised by the integration of services at the appropriate network and user level and by advanced communication features making service provision user-friendly, efficient, and economically sound.

Information Objects: Information objects model conceptual and physical telecommunications resources from the environment under study as well as their relationships for the purpose of management. Depending on its characteristics (dynamicity, visibility, and privacy of the information held by the information object) a given information object is implemented either as a managed object [X.720] or as a directory object [X.500].

Inter-Domain Management: Management activities across a management interface between any two separate domains. The management activities at this interface include the exchange of management information to support locally the end-to-end multi-domain management of a telecommunications service.

Intra-Domain Management: Management activities concerned with the management of resources locally within a domain. Intra-domain management activities and systems are not visible or accessible from outside the domain.

Management Information Model: The set of interrelated management information objects modelling a set of telecommunications resources for the purpose of telecommunications management. (Note that the resources exist independently of their need to be managed.)

Management Information Modelling: The process of defining a management information model.

Management Interface: A TMN management interface is an interoperable interface defining a protocol suite and the messages carried by the protocol. Transaction-oriented interoperable interfaces are based upon an object-oriented view of the communication and therefore all the messages exchanged through the interface deal with object manipulations.

Management Platform: An integrated set of software components providing the necessary tools for developing and running telecommunications management applications. For a list of management platforms used by PREPARE see sections 2.7.2 and 2.8.3.

Management Service: A management service is a set of management functions that is made available by the provider of a service to customers and users of that service. A management service is therefore linked to a specific service offering.

Management System: A management system performs telecommunications management. When a management system conforms to the TMN recommendations it is termed a TMN building block or a TMN.

Multi-Domain Management: Management activities required for the coordinated end-to-end management of a telecommunications service spanning several domains. Multi-domain management comprises the management activities at all the inter-domain management interfaces involved in the management activities for any one service.

Multimedia Teleservice: Teleservice offering multimedia capabilities. The multimedia teleservices used in PREPARE are a multimedia teleconference service and a multimedia mail service.

Network Management: Management of individual telecommunications networks.

One-Stop Shopping: A single point of contact feature allowing customers to buy a complex telecommunications service involving different service and/or network providers through a single commercial and administrative interface. For instance, a global voice VPN service offering could take care of telephone line orders to different network providers on behalf the customer. One-stop shopping can be associated with one-stop billing and one-stop complaining features.

Organisation: Any entity having a juridical existence e.g., administrations, enterprises, universities.

Organisational Situation: An organisational situation is a notional real-world situation of multiple organisations being stakeholders in a set of telecommunications and management services. Rather than being general, as the enterprise models are for the design of generic service management solutions, an organisational situation defines a precise set of stakeholders and defines choices when multiple valid options exist. In order to establish a precise focus for the project's work, PREPARE defined organisational situations as the basis for requirements definition and further design of inter-domain management solutions.

Outsourcing: A business policy where organisations subcontract activities which are not part of their core business to a third party (e.g., operation of the corporate telecommunications network for a car manufacturer).

Peer Cooperation: Management cooperation between two organisations where each organisation utilises management capabilities provided by the other one. For instance, two national ATM service providers exchange the necessary management information in order to provide pan-European services to their customers.

PREPARE (PREPilot in Advanced REsource management): R&D project from the CEC RACE II programme. This book has been written by the PREPARE project.

Private Network Operator (PrNO): A private network operator is an organisation operating a network offering one or more bearer services. A private network operator does not have a legal supply obligation to the general public.

Public Network Operator (PuNO): A public network operator is an organisation which operates a network offering one or more bearer services. A public network operator is legally obliged to provide one or more specific bearer services to the general public.

Role: When describing interactions between stakeholders in a multi-domain management context the concept of role is used to define the allocation of responsibility to individual stakeholders. Responsibilities are thought of as being held by individual agents of the stakeholders and the role concept is used to refine the stakeholder descriptions into smaller, more manageable entities. When an agent of a stakeholder takes on a role in the multi-domain management context it is said to assume a certain responsibility in that context. The responsibility is the basis for the specification of that role (see also: Role Specification).

Role Specification: A role specification is the detailed specification of a role. This comprises the definition of the responsibility held by the role, which is refined into a set of obligations held by that role in order to fulfil its responsibility. Obligations specify what the role must do to fulfil its responsibility. By further refinement, activities are defined specifying how the obligations are to be discharged. Activities are specified as operations by the role on a set of resources. In order to perform its activities the role must have the necessary rights over the resources affected. Thus, a role specification comprises a specification of responsibility, obligations, activities, and rights.

Service Management: Management of customer services.

Stakeholder: A stakeholder is an organisation which defines requirements on a service, or which is responsible for fulfilling another stakeholder's requirements on a service.

Telecommunications Resource: Physical or logical resource within a telecommunications environment (network or service): a network node, a line, a log file, etc.

Teleservice: Telecommunications service that provides a complete capability, including application functions, for communication between users according to protocols established by agreement between network operators and/or service providers. A teleservice covers all seven OSI layers.

Testbed/Broadband Network Testbed: The set of interconnected broadband networks used for the purpose of validating, exercising, and demonstrating PREPARE's inter-domain management solutions. PREPARE used two testbeds, one for each of the two project phases.

TMN (Telecommunications Management Network): In the context of this book, a TMN is a network of interoperable management systems where the means of interoperability between the constituent management systems are in accordance with the principles defined in the ITU recommendation M.3010.

Value-Added Services: Services in which the value-added service provider adds capabilities to services provided by another service provider (i.e., bearer or other value-added services).

VPN (Virtual Private Network): Simulation of a private network based on an underlying shared infrastructure (e.g., a voice VPN based on the public telephone network). Currently, VPNs simulate voice and data networks. Multimedia VPNs based on broadband technologies are foreseen.

VPN Service: A value-added service providing VPN instances to its customers. In the context of PREPARE, VPN was chosen as an example of a value-added service requiring advanced inter-domain management features. VPN services were defined, implemented, and demonstrated by PREPARE.

X Interface: An X interface implements a (set of) TMN x reference points. The x reference points are located between the OS function blocks in different TMNs [M3010/95]. X interfaces were a main focus of PREPARE.

Acronyms

AD	Adjunct
AAL	ATM Adaptation Layer
ACID	Atomicity, Consistency, Isolation, Durability
ACTS	Advanced Communications Technologies and Services
AET	Application Entity Title
API	Application Programming Interface
ASDC	Abstract Service Definition Convention
ASN.1	Abstract Syntax Notation 1
ATM	Asynchronous Transfer Mode
BPM	Business Process Model
CBDS	Connectionless Broadband Data Service
CBR	Constant Bit Rate
CCAF	Call Control Agent Function
CCF	Call Control Function
CCR	Commitment, Concurrency and Recovery
CEC	Commission of the European Communities
CCITT	Comité Consultatif International de Télégraphique et Téléphonique
CMIP	Common Management Information Protocol
CMIS	Common Management Information Service
CMISE	Common Management Information Service Element
CORBA	Common Object Request Broker Architecture
COSS	Common Object Service Specifications
CPN	Customer Premises Network
CSM	Customer Service Manager
DAF	Directory Access Function
DAP	Directory Access Protocol
DCF	Data Communication Function
DIB	Directory Information Base
DIT	Directory Information Tree
DO	Directory Object
DPE	Distributed Processing Environment
DQDB	Distributed Queue Dual Bus
DSA	Directory System Agent
DSF	Directory System Function
DUA	Directory User Agent
ESPRIT	European Strategic Programme for R&D in Information Technology

ETSI	European Telecommunications Standards Institute
EU	European Union
EURESCOM	EUropean institute for RESearch and Strategic Studies in teleCOMunications
FTAM	File Transfer Access and Management
GDIO	Guidelines for the Definition of Information Objects
GDMO	Guidelines for the Definition of Managed Objects
GS	Global Store
GUI	Graphical User Interface
IBC	Integrated Broadband Communications
IDL	Interface Definition Language
IDMIB	Inter-Domain Management Information Base
IDMIS	Inter-Domain Management Information Service
IEC	International Electrotechnical Commission
IEEE	Institute of Electrical and Electronics Engineers
IN	Intelligent Network
IO	Information Object
ISO	International Organization for Standardization
ISP	International Standardised Profile
IP	Internet Protocol
IP	Intelligent Peripheral (Appendix D only)
IT	Information Technology
ITU	International Telecommunication Union
ITU-T	Telecommunication Standardization Sector of ITU
Kbps	Kilobits per second
LAN	Local Area Network
LLA	Logical Layered Architecture
LME	Layer Management Entity
MAF	Management Application Function
MAN	Metropolitan Area Network
Mbps	Megabits per second
MCF	Message Communication Function
MCR	Multicast Router
MF	Mediation Function
MIB	Management Information Base
MMC	Multimedia Conferencing
MMM	Multimedia Mail
MO	Managed Object
MSF	Management Systems Framework

MUX	Multiplexer
NEF	Network Element Function
NMF	Network Management Forum
N_OS	Network Operations System
N_OSF	Network Operations System Function
ODM	Open Distributed Management
ODMA	Open Distributed Management Architecture
ODP	Open Distributed Processing
OID	Object Identifier
OMA	Object Management Architecture
OMG	Object Management Group
OMT	Object Modelling Technique
ONA	Open Network Architecture
ONP	Open Network Provisioning
ORB	Object Request Broker
ORDIT	Organisational Requirements Definition for Information Technology
OS	Operations System
OSF	Operations System Function
OSI	Open Systems Interconnection
PN	Public Network
PNP	Private Numbering Plan
PREPARE	PREPilot in Advanced REsource management
PSA	Provider Service Administrator
PrNO	Private Network Operator
PuNO	Public Network Operator
PVC	Permanent Virtual Circuit
QAF	Q Adaptor Function
QoS	Quality of Service
RACE	Research and development in Advanced Communications technologies in Europe
R&D	Research and Development
RDN	Relative Distinguished Name
RECAP	Requirements Capture
RM-ODP	Reference Model of Open Distributed Processing
SCEF	Service Creation Environment Function
SCEP	Service Creation Environment Point
SCF	Service Creation Function
SDF	Service Data Function
SDH	Synchronous Digital Hierarchy

SF	Security Function
SLA	Service Level Agreement
SMAE	Systems Management Application Entity
SMAF	Service Management Access Function
SMART	Service Management Automation and Re-engineering Team
SMF	Service Management Function
SMI	Structure of Management Information
SMP	Service Management Point
SNMP	Simple Network Management Protocol
S_OS	Service Operations System
S_OSF	Service Operations System Function
SPIRIT	Service Provider Integrated Requirements for Information Technologies
SRF	Specialised Resource Function
SSF	Service Switching Function
TCMO	Transaction Control Managed Object
TDS	Test Design Specification
TINA	Telecommunications Information Networking Architecture
TINA-C	TINA Consortium
TMFA	Telecommunications Management Functional Area
TMN	Telecommunications Management Network
TP	Transaction Processing
UISF	User Interface Support Function
UNI	User Network Interface
VASP	Value-Added Service Provider
VP	Virtual Path
VPI	Virtual Path Identifier
VPN	Virtual Private Network
WAN	Wide Area Network
WSF	Workstation Function
XC	Cross-Connect
XoJIDM	X/Open-NMF Joint Inter-Domain Management working group

Index

Springer-Verlag
and the Environment

We at Springer-Verlag firmly believe that an international science publisher has a special obligation to the environment, and our corporate policies consistently reflect this conviction.

We also expect our business partners – paper mills, printers, packaging manufacturers, etc. – to commit themselves to using environmentally friendly materials and production processes.

The paper in this book is made from low- or no-chlorine pulp and is acid free, in conformance with international standards for paper permanency.

Lecture Notes in Computer Science

For information about Vols. 1–1059

please contact your bookseller or Springer-Verlag